14-120-978

PUBLIC SAFETY

A Growing Factor in Modern Design

A SYMPOSIUM HELD AT THE
FIFTH ANNUAL MEETING
OF THE
NATIONAL ACADEMY OF ENGINEERING

National Academy of Engineering
Washington, D.C.
1970

Standard Book Number 309-01752-1

Available from

Printing and Publishing Office
National Academy of Sciences
2101 Constitution Avenue, N. W.
Washington, D. C. 20418

Library of Congress Catalog Card Number: 70-602959

FOREWORD

Engineers have long recognized the influence of their works on the well-being of society. They have also recognized the continuing responsibility they have for the safety of the public, those who utilize their works. This concern of engineers for the public interest is evidenced by the many actions taken by them as individuals or as members of engineering professional societies. Typical of the attention given to questions of public interest is that involved in the preparation of codes and standards for many different types of structures and devices.

Public safety is of growing importance, particularly in the design of engineering systems having potential for simultaneous hazards to large numbers of people. These systems might be divided grossly into three groups:

- Integrated systems serving large groups of people where a fault can endanger all users (e.g., city water-supply systems, drug manufacturing and distribution systems, food-processing systems, and large planes and ships).
- Widespread systems serving a large number of independent users where the interaction of users is large enough to endanger the lives of many people (e.g., highway and city traffic systems, private and commercial intercity air-traffic systems, airports, hunting and other sports, use of insecticides, and use of detergents).
- Systems utilizing concentrated sources of power with potential for sudden or long-term release of forces or pollutants that can endanger many people (e.g., large fossil-fuel power plants, nuclear power plants, nuclear explosives for peaceful uses, chemical plants, chemical-transport vehicles, and large dams).

It is clear that not all systems or devices have the same potential or the

iii

same probability for damage to large numbers of people. Also, not all systems benefit society equally or in the same way. In addition, in some systems the users accept the risks knowingly, and in others the risks are imposed without the knowledge of the people involved.

With some systems, we have a great deal of data about the probabilities of certain types of consequences, and with others we know only of the potential consequences, with little experience to provide the data on the probability of occurrence. The data on failures seem to be sparsest where the potential consequences are largest, but this is not surprising, for it is exactly these types of devices and structures to which engineers give their greatest attention.

In the design of any system involving the safety of the public, several questions loom uppermost: How is the public interest to be represented in design? How does the designer assess the degree of public safety needed? What guidelines does he have or can he afford to use? Let us look at each of these questions briefly.

How is the public interest represented in design? Two answers might be provided for this question. The first relates to government regulation based on the technical recommendations of professional societies, perhaps in the form of codes of various types. The second involves the economic pressures and surveillance of insurance companies. But in both of these cases we find there are a number of questions when we talk about large systems. For example, under governmental regulation, we find that regulations cannot anticipate all aspects of all problems. The need for regulation is sometimes not recognized soon enough. New technological problems develop before regulatory bodies can be organized. Development of codes takes time. Once in a while we find the government both promoting and trying to regulate the same activity, and this leads to conflicts within government regulation. Of course, with insurance companies, the problems are very similar, particularly when one talks about trying to develop rates based on data.

In the nuclear power industry, for example, the need for a viable regulatory mechanism was recognized early, and appropriate steps were taken to establish such a mechanism. This has been done in a number of other areas as well. In most of these areas it has been found that individual advisory groups have been very valuable in bringing about independent representation of public interest. But the same degree of effort has not occurred in all situations where the potential for damage is or may be as great.

It appears that our professional societies need to broaden the scope of their activities to encompass the larger problems of public interest and that they also need to devote more active attention to these types of questions. We may want to put our local professional groups to use in providing advisory sources to state and local governments.

How does the designer assess the degree of public safety needed? What guidelines does he have or can he afford to use? These questions are perhaps even more important than the first. How do we decide how much safety we need? How much risk can and will the public accept? Basically, how do we balance the benefits and the hazards to society? This is a complex matter because it involves emotional attachments and a different willingness on the part of the public to accept different types of risks.

We as a people accept with some degree of indifference the loss of tens of thousands of lives and millions of dollars of property each year on our highways. Yet would we accept such a risk from our power plants, nuclear or nonnuclear? from our commercial transport systems? from structural failures of bridges? from processed-food poisoning? from building-elevator failure? Each of the designers involved must make assessments of the risks versus the gains. Of course, the public is not conditioned to such levels of loss, nor would we expect them to be. Yet in each case we find that the designer is faced with making the risk-versus-gain assessment.

The papers presented at this symposium make clear, nevertheless, that provisions for public safety are a basic part of every major engineering undertaking and that they have received continuing attention.

The problems are complex, and often the public is not aware of what is being done by engineers in behalf of public safety. It is clear that engineers must improve their communication with the public not only to familiarize it with what is being done but also to obtain public awareness of the issues involved. For example, in high-speed expressways in congested areas, it is sometimes better to remove entrances and exits to improve safety rather than to add more of them; this action may not always meet with public approval, especially where access to businesses is involved.

The Program Committee wishes to express its appreciation to each of the participants in this symposium. The papers presented not only review past practices and recent developments in evaluating the benefits and hazards to society in engineering undertakings but also point the way to more enlightened and broader treatment of all aspects of public safety.

Nunzio J. Palladino
Chairman, Program Committee

Members of the Program Committee,
1969 Annual Meeting of the NAE

Nunzio J. Palladino, *Chairman*	Louis H. Roddis, Jr.
William J. Hall	Philip N. Ross
William C. Mentzer	Chauncey Starr

CONTENTS

Introduction   1
  Eric A. Walker

An Overview of the Problems of Public Safety   3
  Chauncey Starr

Beyond the Law and the Lab in Search of
Public-Safety Design Criteria   16
  James M. Brown

Product Safety—A New Standard of
Excellence for American Industry   24
  David E. A. Carson

Transportation of Hazardous Materials   29
  William K. Byrd

Is Traffic Safety in the Public Interest? When? How Much?   36
  Martin Wohl

A View of Safety in the Petroleum Industry   65
  Donald L. Katz

The Close Relationship between Employee Safety and
Public Safety in Modern Chemical Plant Design   77
  J. Sharp Queener

Safety and Reliability of Large Engineering Structures   82
  Alfred M. Freudenthal

Civil Structures and Public Safety:
Safety in Design and Construction of Civil Structures   88
  John A. Blume

Providing for Public Safety in the
Nuclear Industry—The Engineering Approach   96
  James T. Ramey

Reactor Safety: The Carrot or the Stick?   106
  F. Reginald Farmer

Closing Remarks   113
  Eric A. Walker

Participants   115

INTRODUCTION

ERIC A. WALKER
President, NATIONAL ACADEMY OF ENGINEERING

Today's marvelous technological developments have lightened our work, sometimes speeded our travel, and enhanced our leisure time. Yet every new invention introduces new potential hazards to the public, and it is the task of the engineer to weigh the social benefits that are inherent in these innovations against their possible danger to society. This is something that we as engineers have done for a good many years. But some of the systems we are getting involved in today introduce far bigger questions of public safety than we faced in the past. This is sometimes due to their larger size, their unusual potency, or their more extensive use.

I had my attention called recently to a newspaper article in the Detroit *Free Press*, which dealt with the growing number of disasters triggered by the derailment of railroad cars, usually tank cars filled with volatile chemicals or explosives or something of that sort. Speaking about this, a government safety expert was quoted as saying, "In the past it wasn't a big deal, but now if you have a derailment, you may blow a whole town off the map." In this same article, John Reed, who is a member of the National Transportation Safety Board, said, "It is now obvious that railroad derailment accidents, dangerous enough in themselves, have acquired a new and catastrophic potential for death and destruction previously unknown."

This is only one area in which the industrial activities necessary to meet public needs have introduced questions of public safety of new dimensions, and these questions become particularly acute when the assumption of risk does not involve an overt choice by the individuals involved. We have seen in recent months a number of examples of accidents involving risks of this type where many lives were endangered or lost. These included not only loss of life by chemicals released from derailed tank cars, but also the loss of life in homes demolished by airplanes and property damage and ecological damage brought

1

about by runaway undersea oil wells or broken oil pipelines. If you go back a few more months, we can pick up many more examples that made our headlines, including the failure of bridges, dams, and other civil structures.

But more important than the situations we read about in the papers are the new situations involving more extensive public risk, which we as engineers are beginning to face in increasing numbers. These range from systems in which the potential consequences from certain accidents could be quite severe, but for which the probabilities of failures can be made quite low, to systems having lesser accident consequences, but which have a higher probability of occurrence if they are to be economically feasible.

Our purpose here today is to give attention to the knowledge, methods, and practices that we have built up over the years in maintaining a balance between the benefits and hazards to society inherent in such systems. We hope to shed light on the extent to which further guidelines are needed in assessing the degree of permissible risk.

As an organization of engineers, the National Academy of Engineering recognizes both the difficulty of this task and its importance to the well-being of society. We hope that this symposium will aid the Academy in defining its role in public safety on the national scene.

AN OVERVIEW OF THE PROBLEMS
OF PUBLIC SAFETY

CHAUNCEY STARR

The design of technological systems always involves an overt choice among alternative synthesized solutions and operating conditions in order to optimize eventual performance. The criteria for such an optimization are usually well defined for most components, "black-boxes," or subsystems in already specified large systems. One of the most common criteria is, of course, direct monetary cost. Technological analyses for disclosing the relationship between expected performance and monetary costs are a traditional part of all engineering planning and design. The inclusion in large system studies of *all* societal "costs" (indirect as well as direct) is much less customary and obviously makes the analysis more difficult and less certain quantitatively.

Of obvious interest is the social "cost" associated with the public-safety consequences of design decisions. For example, if analysis of a new development predicts an increased annual income of x percent, but also predicts an associated accident risk of y fatalities annually, then how are these to be compared in their effect on the "quality of life"? Because the penalties or risks to the public arising from a new development can always be reduced by applying constraints, there will usually be a functional relationship (or trade-off) between utility and risk, the x and y of our example.

We have many historical illustrations of such trade-off relationships empirically determined. For example, automobile and airplane safety has been weighed continuously by society against economic costs and operating performance. In these and other cases the real trade-off process is actually one of dynamic adjustment, with the time behavior of many portions of our social systems out of phase due to the many separate "time-constants" involved. The ready availability of historical accident and health data for a variety of public activities provides an enticing quantitative stepping-stone to an evaluation of this particular type of social cost. The corresponding social benefits arising

3

from some of these activities can be roughly determined. *On the assumption that such historical situations have achieved a socially acceptable and reasonably optimum trade-off of values, any generalizations developed might then be used for predictive purposes.* This approach could give a rough answer to the deceptively simple question, "How safe is safe enough?"

The pertinence of this question to all of us, and particularly to government regulatory agencies, is obvious. Hopefully, a functional answer might provide a basis for establishing performance "design objectives" appropriate for technology entering into social use.

Voluntary and Involuntary Activities

Societal activities fall into two general categories, those in which the individual participates on a "voluntary" basis and those in which the participation is "involuntary," i.e., imposed by the society in which the individual lives. The process of empirical optimization of benefits and costs is fundamentally similar in both cases—namely, a reversible exploration of available options—but the time required for empirical adjustments (the time-constant of the system) and the criteria for optimization are quite different in the two situations.

In the case of voluntary activities, the individual uses his own value system to evaluate his own experiences. Although his eventual trade-off may not be consciously or analytically determined or based upon objective knowledge, it is nevertheless likely to represent, for that individual, a crude optimization appropriate to his value system. For example, a city-dweller may move to the suburbs because of a lower crime rate and better schools, at the expense of increased highway travel time and accident probabilities. If, subsequently, the traffic density increases, he may decide the penalties are too great and move back to the city. Such an individual optimization process can be comparatively rapid (because of the rapid feedback of experience to the individual), so that the statistical pattern of a large social group may be an important "real-time" indicator of societal trade-offs and values.

Involuntary activities differ in that the criteria and options are determined not by the individuals affected but by a controlling body. Such control may be in the hands of a government agency, a political entity, a leadership group, an assembly of authorities or "opinion-makers," or a combination of such bodies. Because of the complexity of large societies, only the control group is likely to be fully aware of all the criteria and options involved in the decision process.

Further, the time required for the feedback of the empirical experience resulting from the controlling decisions is apt to be very long. The feedback of cumulative individual experiences into societal communication channels (usually political or economic) is a slow process, as is the process of altering the direction of the control group. We have many examples of such "involuntary" activities, with war being perhaps the most extreme case of the operational separation of the decision-making group from those most affected.

In examining the historical benefit–risk relationships for involuntary activities, it is important to recognize the perturbing role of public psychological acceptance of risk arising from the influence of authorities or traditional dogma. Because in this situation the decision-making is separated from the affected individual, society has generally clothed many of its controlling groups in an almost impenetrable mantle of authoritative wisdom. The public generally assumes that the decision-making process is based on a rational social benefit-risk analysis. While it often is, we have all seen disclosed after the fact examples of irrationality. It is therefore important to omit such "witch-doctor" situations from the selection of examples of optimized involuntary activities— because, in fact, they are not yet optimized but are only in the initial stages of option exploration.

Quantitative Correlations

With this description of the problem, and the associated caveats, we are in a position to discuss the quantitative correlations. For the sake of simplicity in this initial study, I have taken as a measure of the physical risk to the individual the fatalities associated with each activity. Although it might be useful to include all injuries (which range from 100 to 1,000 times the number of deaths), the difficulty in obtaining data and the unequal significance of varying disabilities made it an inconvenient complexity. So the risk measure used here is the statistical probability of fatalities per hour of exposure of the individual to the activity considered.

The choice of the "hour's exposure" unit was deemed to be more related to the individual's intuitive process in choosing an activity than an "annual" unit, and using the annual unit did not appear to change the substance of the results. Another possible alternative, the risk per activity, involved the comparison of too many unlike units of measure. Thus, in comparing the various transportation modes, one could use risk per hour, per mile, or per trip. As this study was directed toward exploring a methodology for determining social acceptance of risk, rather than the safest mode for a particular trip, the simplest common unit of risk—risk per hour of exposure—was chosen. A future study might explore this issue further.

The social benefit derived from each activity was converted into a dollar equivalent, as a measure of integrated value to the individual. This is perhaps the most uncertain aspect of the correlations because it reduced the quality-of-life benefits of an activity to an overly simplistic measure. Nevertheless, the correlations seemed useful, and no better measure was available. In the case of the voluntary activities, the amount of money spent on the activity by the average individual involved was assumed to be proportional to its benefit to him. In the case of the involuntary activities, the contribution of the activity to the individual's annual income (or the equivalent) was assumed to be proportional to its benefit. This assumption of a roughly constant relationship between benefits and expenditures, for each class of activities, is clearly an approximation. However, as we are dealing in orders of magnitude, the distortions likely to be introduced by this approximation are relatively small.

In the case of transportation modes, the benefits were equated with the sum of the monetary cost to the passenger and the value of the time saved by that particular mode as compared to a slower competitive mode. Thus, airplanes were compared to automobiles, and automobiles were compared to public transportation or walking. Public-transportation benefits were equated with their cost. In all cases, the benefits were taken on an annual dollar basis because it seemed to be more appropriate to the individual intuitive process. For example, most luxury sports require an investment and upkeep only partially dependent on usage. The associated risks, of course, exist only during the hours of exposure.

Probably the best example of the analysis of an involuntary activity is the benefit and risk associated with the use of electricity. In this case the fatalities included those arising from electrocution, electrically caused fires, the operation of power plants, and the mining of the required fossil fuel. The benefits were estimated from a United Nations study of the relationship between energy consumption and national income, and the fraction associated with electric power was used.

Compared to the use of electricity in industry, the more subtle contributions of electric power to our quality of life are, of course, less easily evaluated. For instance, the availability of refrigeration has certainly improved our national health and the quality of dining, and the electric light has certainly provided great flexibility in living. The contributions of television may be more uncertain, but the public response indicates that it is a positive element in our living patterns. Perhaps, however, the income measure used here is sufficient for the present purpose.

Information on voluntary risk acceptance by individuals as a function of income benefits is not easily available, although we know that such a relationship must exist. Of particular interest therefore is the special case of miners exposed to high occupational risks. Figure 1 is a plot of the accident rate and severity rate of mining injuries versus the hourly wage. The acceptance of individual risk is an exponential function of the wage and can be roughly approximated by a cube relationship in this range. If this relationship has validity, it

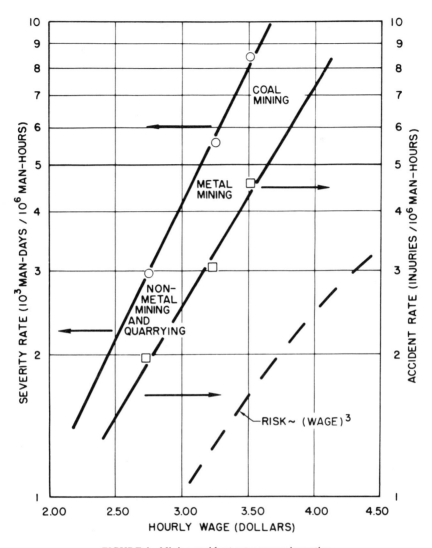

FIGURE 1 Mining accident rates versus incentive.

may mean that three parameters in the quality of life (perhaps health, essentials, and recreation) are partly influenced by any increase in available personal resources, and thus the increased risk acceptance is proportionally motivated. The degree to which this relationship is voluntary for the miners is not obvious, but it is interesting nevertheless.

The results for those societal activities studied, both voluntary and involun-

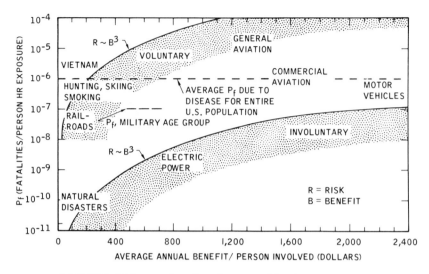

FIGURE 2 Risk versus benefit, voluntary and involuntary exposure.

tary, are assembled in Figure 2. Also shown in Figure 2 is the cube relationship between the risk and benefit characteristics of Figure 1. For comparison purposes, the average risk of death from accident and disease is also shown. As the average accident fatalities are only about one-tenth that of disease, their inclusion is not significant.

Risk Comparisons

Several major features of the benefit–risk relations are apparent, the most obvious being the *separation by several orders of magnitude between the voluntary and involuntary societal acceptance of risk.* As one would expect, we are loath to let others do unto us what we happily do to ourselves.

The disease death rate appears to play a yardstick role in determining the acceptability of risk on a voluntary basis. Most sporting activities are surprisingly close to the disease level—almost as though the individual's subconscious computer adjusted his sporting courage to meet but not exceed the statistical mortality due to involuntary exposure. Perhaps this defines the demarcation between boldness and foolhardiness.

The Vietnam war statistic is shown because it raises an interesting point. Its risk is only slightly above the average disease risk. Assuming that some long-range societal benefit was anticipated from this war, the related risk as seen by society as a whole is not substantially different from the average nonmilitary disease risk. However, to the exposed military-age group (20–30) the Vietnam risk is about ten times the normal mortality (accident plus disease) rate for that age. Hence, a difference in perspective between the population as a whole and those directly exposed. This raises the question of whether the disease risk pertinent to the average age of the involved group might not make a more meaningful comparison than the national average. This would complicate these simple comparisons, but it may be more significant as a yardstick.

The risk positions of general aviation, commercial aviation, and motor vehicles deserve special comment. Motor vehicles originated as a voluntary sport and have had a half century to become an essential utility. General aviation is still highly voluntary. Commercial aviation is partly voluntary and partly essential and additionally is subject to government administration as a transportation utility.

The motor vehicle has now reached a mature benefit–risk balance, as shown in Figure 3. It is interesting that its present risk level is only slightly below the

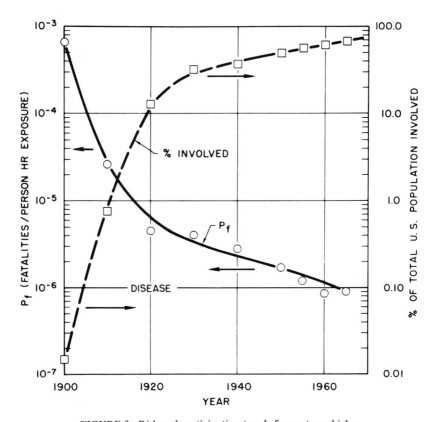

FIGURE 3   Risk and participation trends for motor vehicles.

basic disease level. In view of the high percentage of the population involved, this probably represents a true societal judgment on the acceptability of risk in relation to benefit. It also appears from Figure 3 that future reductions in the risk level will be slow in coming, even if the historical trend of improvement can be maintained.

Commercial aviation has barely approached a risk level comparable to that set by disease. The trend is similar to that of motor vehicles, as shown in Figure 4. However, the percentage of the population participating is now only a

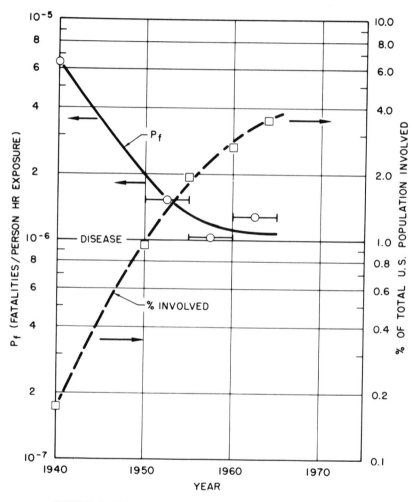

FIGURE 4   Risk and participation trends for certified air carriers.

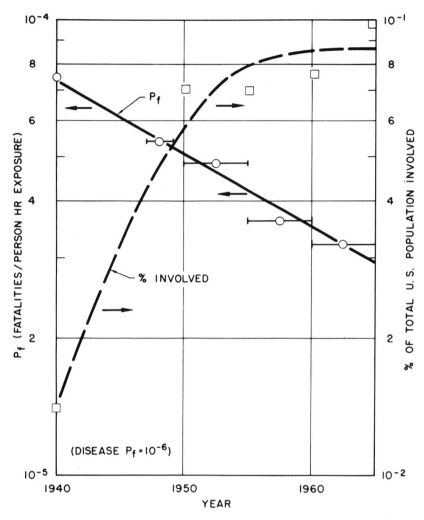

FIGURE 5   Risk and participation trends for general aviation.

twentieth that of motor vehicles. Increased public participation will undoubt-
edly increase the pressure to reduce the risk, because for the general population the benefits are much less than those associated with motor vehicles.
Commercial aviation has not yet reached a mature optimum of benefit–risk
trade-off.

General aviation has similar trends, as shown in Figure 5. Here the risk
levels are so high (twenty times the disease risks) that this activity must prop-

erly be considered in the category of an adventuresome sport. However, the rate of risk is decreasing so rapidly that eventually the risk for general aviation may approach commercial aviation in performance. Since the percentage of the population involved is very small, it appears that the present average risk levels are acceptable to only a limited group.

The similarity of the trends in Figures 3, 4, and 5 may be the basis for another hypothesis: *The acceptable risk is inversely related to the number of people participating in an activity.*

The product of the risk and percentage of the population involved in each of the activities in Figures 3, 4, and 5 has been plotted in Figure 6. This graph represents the historical trend of total fatalities per hour of exposure of

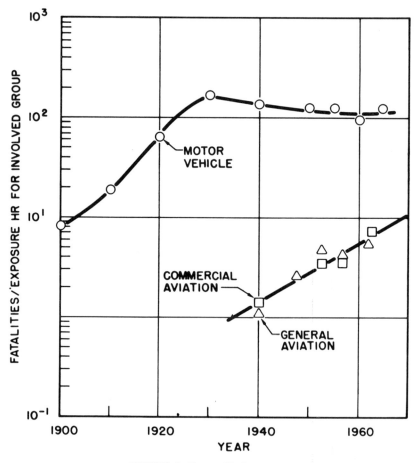

FIGURE 6   Group risk, by year.

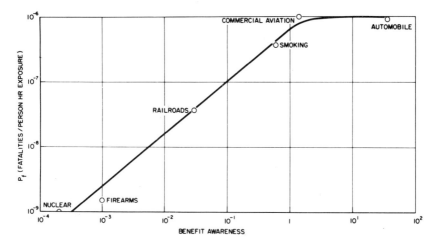

FIGURE 7   Accepted risk versus benefit awareness.

the population involved. The leveling-off of motor-vehicle risk at about 100 fatalities per hour of exposure of the participating population may be significant. Because most of the U.S. population is involved, this rate of fatalities may have sufficient public visibility to set a level of social acceptability. It is interesting, and disconcerting, that the trend of fatalities in aviation, both commercial and general, is uniformly upward.

Public Awareness

As a final attempt to probe our societal attitudes quantitatively, I related these risk data to a crude measure of public awareness of the social benefits (Figure 7). The benefit awareness was arbitrarily defined as the product of the relative level of advertising, the square of the percentage of population involved in the activity, and the relative utility (or importance) of the activity to the individual. Perhaps these assumptions are too crude, but Figure 7 does support the reasonable position that advertising the virtues of an activity increases the public acceptance of a greater level of risk. This, of course, could subtly produce a fictitious benefit–risk ratio—as may be the case for smoking.

Atomic Power Safety Example

While we recognize the uncertainty inherent in the quantitative approach discussed in this presentation, the trends and magnitudes may nevertheless be of sufficient validity to warrant their use in determining national "design objectives" for technological activities. How would this be done?

Let us consider as an example the introduction of nuclear-power plants as a principal source of electric power. This is an especially good example because the technology has been nurtured, guided, and regulated primarily by government—with industry undertaking the engineering development and diffusion into public use. The government specifically maintains responsibility for public safety. Further, the engineering of nuclear plants permits continuous reduction of accident probabilities with an associated substantial increase in cost. Thus the trade-off of utility and potential risk can be made quantitatively.

Another feature of the nuclear power plant is that the historical empirical approach to achieving an optimum benefit–risk trade-off is not pragmatically feasible. All such plants are now so safe that it may be 30 years or longer before meaningful risk experience will be accumulated. By that time, many plants of varied design will be in existence, and the empirical accident data may not be applicable to those being built. So a very real need exists now to establish design objectives on a predictive-performance basis.

Let us first arbitrarily assume that nuclear power plants should be twice as safe as coal-burning plants so as not to increase public risk. Figure 2 indicates that the total electric-power risk is about 2×10^{-9} fatalities per person per hour of exposure. Fossil-fuel plants contribute about one-fifth of this, so we target nuclear plants at one-tenth of this risk. Assuming continuous operation, the nuclear plant would have to achieve a fatality level of not more than two deaths per million population per year ($2 \times 10^{-9} \times 10^{-1} \times 10^6 \times 10^4$ hr/yr = 2). In a modern society, a million people may require a million kilowatts of power, and this is about the size of most new power stations. So we now have a target risk limit of two deaths per year per 10^6-kW power station.

Technical studies of the consequences of hypothetical extreme (and unlikely) nuclear-power-plant catastrophes that would disperse radioactivity into populated areas have indicated that about 10 lethal cancers per million population might result. On this basis, each such power plant might statistically have one such accident every 5 years and meet the risk limit set. However, such a catastrophe would completely destroy a major portion of the nuclear section of the plant and would require either complete dismantling or years of costly reconstruction. Because power companies expect plants to last about 30 years, the economic consequences of a catastrophe every 5 years would be completely unacceptable. In fact, the operating companies would not accept one such failure, on a statistical basis, during the normal plant lifetime.

It is likely that, in order to meet the economic performance requirements of the power companies, a catastrophe rate of less than one in about 100 plant-years would be needed. This would be a public risk of 10 deaths per 100 plant-years, or 0.1 death per year per million population. *So, the economic investment criteria of the nuclear-plant user, the power company, would probably set a risk level two hundred times lower than the present socially accepted risk associated with electric power, or forty times lower than the risk associated with present coal-burning plants.*

An obvious design question is whether a nuclear-power plant can be engineered with a predicted performance of less than one catastrophic failure in 100 plant-years of operation. I believe the answer to that question is yes, but that is a subject for a different occasion. The principal point is that the issue of public safety can be focused on a tangible, quantitative engineering design objective.

This example reveals a public-safety consideration that may apply to many other activities. As in this case, the economic requirement for the protection of major capital investments may be a more demanding safety constraint than social acceptability.

Conclusion

The application of this approach to other areas of public responsibility is self-evident. It does provide a useful methodology for answering the question, "How safe is safe enough?" Further, although this study is only exploratory, it does reveal several interesting points.

First, the indications are that the public is willing to accept voluntary risks roughly a thousand times greater than involuntary exposures. Second, the statistical risk set by disease appears to be a psychological yardstick for establishing the level of acceptability of other risks. Third, the acceptability of risk appears to be crudely proportional to the cube of the benefits (real or imagined). Fourth, the social acceptance of risk is directly influenced by public awareness of the benefits of an activity, as determined by advertising, usefulness, and the number of people participating. Fifth, in the application of these criteria to atomic-power-plant safety, it developed that an engineering design objective determined by the economic criteria resulted in a design target risk level about two hundred times lower than the present socially accepted risk for electric power.

Perhaps of greatest interest is that this methodology for revealing existing social preferences and values may be a means of providing the social benefit-cost insight so necessary for judicious national decisions in new technological developments.

BEYOND THE LAW AND THE LAB IN SEARCH OF PUBLIC-SAFETY DESIGN CRITERIA

JAMES M. BROWN

Two years ago, Wilfred Owens of the Brookings Institution wrote a fable on how people solved their transportation problems. His beginning sentences so intrigued me that I am going to paraphrase them in the format of a public-safety orientation.

There once was a nation of 200 million people that was the most technologically advanced country in all the world. At the national level they had an abundance of public-protection systems, techniques, and institutions, but at the local level these often turned out to be quite ineffective in preventing accidents. And as luck would have it, all accidents occurred at the local level. . . .

As the scientists created wonder after wonder for the benefit of the people, the people began to wonder why it was that some of the wondrous benefits carried increasing propensities for killing and maiming their fellows and destroying their property. But that aspect was beyond the responsibility of the scientists. The people read about disasters at breakfast day after day, and they began to suffer with indigestion. They listened to broadcasts recounting tragedies on the way home and were irritable at dinner. The late-evening telecasts caused them to toss and turn in their beds. Old words representing places and things took on an ominous significance. And the list of such words grew day by day, as did their unsettling effect. Public authorities became more apprehensive as the list increased and the frowns of the people multiplied. But there was little that public authorities could do about it. Still, on occasion, when they met with their fellows, they would talk in such words and shake their heads. How much should they be concerned, they wondered, over Texas City; South Amboy; Port Chicago; Roseburg, Oregon; Richmond, Indiana; the Searcy silo; and the Apollo capsule incidents? Should they anticipate further events such as Palomares, the Greenland Ice Cap, Chalk River, Windscale, SL-1, the PRCD incident? Would Tacoma Narrows or Silver Bridge, or the MTC

602 and the Wychem barge, or the *Torrey Canyon* or Santa Barbara happen to them? What could they do to avoid a South Charleston, West Virginia, or a Newton, Alabama; a Laurel, Mississippi, or a Crete, Nebraska; an Indianapolis Ice Rink or a McCormick Place? Could a Cranberry Thanksgiving, or a Northeast blackout, or a Dugway, Utah, expose to public view their local vulnerability? And day by day the list grew: Norwich, Connecticut; Danbury, Texas; Fairfield, California; Allentown, Wisconsin—was no place immune?

The people began to see connections and repetitions. Donora, London, and Los Angeles had a common link. Thalidomide, Chloromycetin, LSD, "the Pill," and even common aspirin and coffee began to represent only differing degrees of similar apprehensions. Tampico, Mexico, and Columbia both proved what many had suspected: that parathion should not be baked into bread. The chain that included the St. Lawrence seaway, sea lampreys, alewives, coho salmon, and DDT caused a few timid souls to be concerned about the proposed sea-level canal across the isthmus. And, murmured the public officials among themselves, we could even end up buried in our own garbage if a strike lasted very long.

"Do not fear," said some scientists. "What technology can do, technology can undo. We'll make building bricks out of your garbage and turn a liability into an asset."

"But wait," cried the chemists. "Garbage bricks could generate methane gas, and soon the buildings would explode."

The planners and the railroaders marched to the rescue. "Mr. Mayor," they said, "We'll haul your garbage to Pennsylvania in mile-long trains."

"Only over our dead bodies," said the Pennsylvanians, "for if you do we might as well be dead."

"Dump it at sea" directed the mayor in an authoritarian voice.

"You'll ruin our beaches," shouted the bathers at Far Rockaway, waving their ballots angrily.

"You'll upset the marine ecology," screamed the conservationists, as they dialed their congressman.

"Then burn it up," the mayor ordered.

"But we are on an air-pollution alert now," cautioned his advisers.

"Bury it deep beneath the ground," the mayor growled. But the commissioners sadly informed him that the city's remaining land was already overpromised for public housing.

"Then buy more land!" wailed the mayor, but the budget director and the tax collector sadly shook their heads.

"I'll call the governor," moaned the poor mayor, "No, that won't help; get me through to the President."

The President had troubles of his own, but cautioned the mayor to move cautiously.

"If you could just let us have $10 billion," wheedled the mayor. The President's personal secretary apologized that Congressman Williams had just stepped in to have a chat, but that the President deeply sympathized.

The mayor sat perplexed with thought. Finally a ray of hope began to show on his face. "Miss Jones," he addressed the walnut box on his desk, "come here a moment, please." She came. "Please shut the door," he asked; and then, "When does my term expire?"

And so it went across the land, and the accidents increased, and the troubles grew, and the people became more restive and concerned.

"Let us look abroad for solutions to our problems," some suggested. "Two thousand died at Longerone. What can we learn there?"

A State Department expert spoke to that: "Prevention is the lesson there. The Soviets have charged that the Viaont dam earthslip would have been anticipated by competent engineers."

"But they didn't anticipate the one above Tashkent," a foreign correspondent responded.

A planner and a builder spoke. "In any event, the government rushed to the rescue. They built new plywood homes of modern design in a better location, and set aside a large fund for compensation."

"That was a solution?" snorted a sociologist. "The people didn't like the homes and wouldn't settle in the new location. They wanted to return to their former homesites."

"And by the time everyone had argued about how the money should be distributed, and how to make sure that no one got more than they were entitled to, and which agencies of the government should make the disbursements, and what records should be kept, and committees were appointed and inspections carried out, and studies made, most of the money was gone," added political scientists.

A psychologist commented: "Not only that, but several years had gone by, and the people who needed the money for rehabilitation had rehabilitated themselves, to the extent that they could do so. The people became very angry. They even took up arms and threatened to seize the dam."

"What happened next?" a student asked.

"Well, there was only one dependable solution, as always," said the law professor. "Something had to be done. Such things should not go uncorrected. They sued the engineers who designed the dam, all nine of them. Only eight finally went to trial. One committed suicide."

The student scratched his head. "What good did that do? Even if they were found liable—after all, nine—I mean eight—engineers—and over 2,000 died. And what if they couldn't have predicted what happened?"

"Well," said the professor, "the important thing is to assess who is at fault."

"But what about the people?" asked the student. "Eight engineers—and 2,000 victims. Assessing damages won't do much good, will it?"

"It won't produce much money, if that's what you mean," said the professor, "but that's the chance you take. What it will do is make the engineers think a little more carefully before they build the next dam."

"Maybe they just won't build the next one," suggested the student.

A practicing lawyer stepped up. "Things would have gone differently if it had happened in this country. Under the Tort Claims Act, we'd have sued the government, like we did over Texas City."

A timid Texan far back in the crowd reminded the lawyer that the plaintiffs had lost in that suit, after 8 years in court.

"Only because of the 'discretionary function clause'," the lawyer answered, "and look at the precedent value of the *Dalehite* case. It's been cited over and over since Texas City."

The law professor shook his head. "It's rather unreliable since *Indian Towing*, which some say reversed *Dalehite*."

"But the Supreme Court hasn't said so," said the lawyer.

"No, but they haven't hit the exact issue again, and they have distinguished *Dalehite* into a pretty tight corner," argued the professor.

"Have there been any recent accidents similar to Texas City?" asked the student.

"Well, the South Amboy explosion was quite devastating," said the professor.

"I read about that one," said the student. "They maneuvered for 9 years without even getting to trial, didn't they?"

"No wonder," exclaimed the lawyer. "Just look at the situation. Thirty-one deaths; millions of dollars of property damage; 186 separate lawsuits; 8,000 plaintiffs; 30 defendants, including the federal government; nine different courts in four different states. And the producer of the explosives gone insolvent. These things take time."

"But Judge Forman finally had to arrange an out-of-court settlement, and only raised a measly 3 million dollars," complained the student.

"Shows the strength and flexibility of the legal process," boasted the lawyer.

"Then we'd better burn it down," growled the student. The lawyer was appalled at that.

And the people became more restive and concerned. "Who speaks for us?" they wondered, sometimes aloud. And spokesmen stepped forward, trumpets in hand. "Responsibility is what we need," cried Nader. "Consider the significance of what you are doing," admonished Carsen. "Our earth is small and lonely, and our resources are limited," warned Barbara Wells.

"They do make some sense," the producers admitted privately, "but they are so radical, and really not well informed." And they went on in their customary ways, a bit irritated that they should be told such things. And the accident lists continued to grow. The people said, "Someone should do something," and they looked about, but "someone" was hard to find.

The lawyers magnanimously stepped forth. "We'll sue the engineers," they offered.

"Oh, no, not us," was the reply. "We do our best to prevent any harm. And we put warning signs on anything dangerous. People really should be more careful. Suing us is no remedy."

But the lawyers were warming quickly to the task. "*Respondeat superior* will produce the cash, and look at the trends. Absolute liability is here to stay. *Green* v. *Tobacco* is the rule today."

The student prodded them once more. "It's 11 years since Mr. Green died with his lungs all black from the smoke inside. And the case is up on appeal again. Mrs. Green's death of dissatisfaction will precede her winning that wrongful-death action."

"You must learn to speak like a lawyer," the professor scolded. Just then the law clerk cantered up, an advance sheet clutched in his slender hand. "The 5th Circuit Court has done us in; they've just reversed the *Green* case again."

The lawyer quick-scanned the dissenting opinion. "But they've had to guess at the Florida law; what they say it is, it isn't at all. Well—win a few and lose a few; contingency fees will carry us through." They had worried a bit over Workmen's Comp. And Keeton-O'Connell could give them a whomp. But the SST and pesticides; nuclear power and herbicides; the ABM and insecticides; with drugs and hazardous products galore, all paths still led to the courtroom door.

And the people became more and more restive and concerned. But the lawyers spread balm on their rising fears. "Don't worry, we'll sue all the engineers."

But that didn't seem like much of a solution.

Fortunately, among the people were a few who looked about them and began to wonder if perhaps something could and should be done. The probability chain lottery didn't work too well. The insurance companies were getting a bit cautious. The government was stirring and beginning to frown. The lawyers' games didn't seem to help. And failures in risk management and hazard control were becoming demonstrably unselective in the selection of victims. Sometimes an accident occurred rather close to home.

Somber warnings by competent people began to be taken seriously. Lake Erie really could be turning into a future swamp. It was even conceivable that we could render our environment uninhabitable. Too many things appeared to be getting out of hand, to the detriment of all the people, even the lawyers and the producers, and the scientists. Technological developments seemed to have outpaced the traditional protective mechanisms that society had structured. And finally people from various disciplines began to talk together and to find that their various skills and abilities could be combined to accomplish what none of them could do individually. They discussed the question of what risks could and should be tolerated by society and tried to establish criteria for making sound decisions in that regard. They began to realize that there was in fact a hierarchy of exposure levels. Those who had no knowledge of the risks to which they were exposed and who thus had no option of removing themselves from the area of hazard were felt to be deserving of the greatest degree of protection from risks. But since some of the hazards to which any person might be exposed were within the potentials of products that were of general benefit and greatly desired by society, it was not expedient to eliminate all risks. What could be done in such cases was to combine talents to identify and describe such risks accurately and then to structure defensive mechanisms well designed to mitigate the detrimental consequences of those risk realizations that in fact did occur. One of the problems inherent in coordinating efforts to achieve such a goal was that on occasion the distribution to public-safety organizations of information regarding the best means of combating the hazards inherent in certain products or processes often came to be used to structure legal liability after an accident. But if the processes for providing compensation for victims and for determining the occasions that justified punitive sanctions were separated, and if the system for redistributing the costs of compensation could be synchronized with a mechanism that considered, in a broad context, when recoupment should be assessed and to what degree, this shadow might be eliminated.

With the availability of computers, it has become possible to assess accurately the environment of an impact area exposed to each known hazard so that special controls could be imposed for critical areas while greater risks could be taken across the general domain. Thus, selective exposure could permit a greater capacity for accommodating hazardous products in general. The full understanding of risks and the full dissemination of protective information to those charged with the responsibility for containing and eliminating deprivations that did occur would permit a highly effective system of public protection to be created generally, and this might make it possible to reach the point where no accident could expand into a catastrophe. And if the needs of victims of the failure to manage risks adequately were objectively explored, analyzed, understood, and provided for, in time and kind as well as in funded costs, those who did suffer would not suffer so much, nor would they feel that they had merely been offerings to be sacrificed on the altar of profit.

Even when adequate information is available on product characteristics and impact areas; even when such areas are analyzed for selective degrees of protection; even when the computers can correlate information and "red-flag" those situations that increase the area of possible risk exposure beyond provided-for perimeters, or that expose an area to a type of risk it is not protected against, still serious accidents will happen. To prevent these from ballooning into catastrophes, an effective response mechanism must be built to deal with the crisis situation. Such a response must be built around comprehensive plans for organizational functioning in both pre- and post-accident situations, and such plans have to be widely known, understood, and tested to gain proficiency and to avoid overlaps, duplication, and omission of essential corrective activities. Necessary equipment has to be available, properly stockpiled. Key community installations have to be identified and linked together, and priorities in restorative efforts have to be effectively established; key organizations must be able to respond with trained personnel, properly equipped; communication facilities must be adequate not only to coordinate combative efforts but to permit organizational representatives to transmit detailed and accurate information rapidly back to their headquarters; tasks assigned to various organizations must be geared to their traditional activities and must be sufficiently specific to avoid confusion; a hierarchy of authority must exist within well-ordered lines, and it must operate within a specific overall disaster-prevention plan; and jurisdictional boundaries must be uniformly structured. We should not have to hear of firetrucks being turned back for lack of a pass, or of evacuees being directed through rather than away from a cloud of escaped chlorine gas. The Grandecamp explosion killed 536, but when the High Flyer exploded a few hours later, only one person was killed, because the warning signs were read correctly the second time. Five hundred and thirty-six deaths is too high a price to pay for such an education.

We need to understand more thoroughly the phases of an ongoing deprivation. An explosion not only damages with blast, flying debris, and fire. It also generates secondary deprivational entities. Many were injured at Texas City by flying window glass. Bottom-anchored venetian blinds on the harbor side might have trapped most of these secondary missiles. There are many little things as well as big things that make the difference. The list of things we know but do not apply well is long. But we really know very little, nonetheless, about accidents or catastrophes. We could build a list of things to do that are essential to designing effectively for public safety that would fill a book. But they would eventually be abstractable to a need to learn, a need to disseminate what we have learned, and a need to systematize the employment of that knowledge. To do any of these things, we must reach beyond our own disciplines and work together more effectively. Disaster should not always lurk one trailer truck or tank car beyond our door. In the midst of this second technological revolution, we should be able to place our trust in something more dependable than the hovering specter of tort liability. We can create change, but we have to design for progress. If the price of progress is the taking of a calculated risk, let us put our emphasis on calculation, not on risk. If we are to design for public safety, let us not get so wrapped up in our technical efforts that we forget the public. Accidents do happen only at the local level, and as luck would have it, that's where we live, too.

PRODUCT SAFETY—A NEW STANDARD OF EXCELLENCE FOR AMERICAN INDUSTRY

DAVID E. A. CARSON

Americans have always believed that anything is possible. Advances in technology and engineers' imaginations have been reassuring them in that belief. And on February 20, 1962, they received conclusive proof. At the moment astronaut John Glenn splashed down from his triple orbit of the earth, a new bench mark in safety and precision was reached. The belief that total safety in product and system design was possible became a fact of life for each of us. The American public suddenly knew conclusively that there was no real need for home appliances to short-circuit, for steering columns to be defective, for the inadequacies of highway systems. John Glenn became the ultimate consumer—putting his life confidently in the hands of American design engineers, getting the maximum use and enjoyment of their product, and, having used the item once, putting it in a museum and waiting for a new model to make the next trip an even more spectacular one.

The American consumer began to stretch his muscles, and "consumer power" became a new cliché—and a new challenge for each of us to contend with.

Today's customer-oriented society and the forecasts for tomorrow point to increasing demands that every item and every service be designed to protect the individual consumer from even the possibility of injury, inconvenience, or economic loss. Traditional legal doctrines and defenses are upset daily, shifting the financial cost of accidents from the injured party to designers, manufacturers, or the public at large. The failure to consider the safety of products or systems will more and more be construed as negligence in manufacture and design. The buyer is no longer warned to beware. From here on out the warning will be directed to the seller.

The American consumer has become impatient. He will not allow a return

24

to less sophisticated standards of safety. He will accept no justification for delay and no excuse for failure. Whether design errors are trivial or serious is not the point. The public will continue to rate design factors by the standards of design efficiency inherent in successful voyages through space.

For engineers and designers the mandate is a simple one: Make it better and make it safer.

For the manufacturer and designer alike, the problems are legion. Products and systems must be safe not only for the use intended, but safe even when misused. The demands and the cost of design research increase. The engineer's responsibility for a safe design extends from conception to destruction or disposal in the inevitable junk heap—and each step in between.

When any new product or service is put on the market, the public will judge it only on its ability to do a job better, to make recreation more exciting, travel more convenient, and living a little easier. The public will not judge safety. They assume it. They assume that, for example, new highways are designed for travel at 65 mph (even when posted for 40 mph). They assume that a vast retinue of government agencies, commissions, and bureaus are involved in maintaining safety standards and improving dependability—and they assume correctly.

The complexity and extent of safety considerations for the public is evidenced in every newspaper and on every news telecast: the calling back of thousands of automobiles of a particular model; the damage caused by offshore oil drilling leaks; the tragedy of black lung among Appalachian coal miners; the discussion of safety factors in the consideration of ABM missile sites; the fantastic growth of air traffic over major population centers and the safety implications of massive jetports; the wholly new question of nuclear waste material and the radiation factor in increasing numbers of industrial activities; the estimated 6 to 16 million workers in the United States alone currently threatened with loss of hearing due to unacceptable noise levels on the job; and the increasing numbers of city dwellers threatened with hazardous air pollution well beyond acceptable threshold levels of health tolerance.

These are just a handful of the new and old health and safety problems receiving a hard look from virtually every government department and bureau and from an increasing number of associations, leagues, and task forces. There are the studies of the American Conference of Governmental Industrial Hygienists, ordnance standards set by the Department of Defense, hazardous-material standards of the Department of Transportation, safety and health standards of the Department of Labor, the National Fire Codes of the National Fire Protection Association, the American Safety Standards of the United States of America Standards Institute, the National Commission on Products Safety, the clean air task forces, and of course, the consumer's latest institution, Ralph Nader. The involvement of the insurance industry in each of these areas is intimate, critical, and inescapable. How insurance fits in, and to what extent, is still a question.

The role of the insurance industry has been and continues to be spotlighted from both sides of the product-safety issue, with the manufacturer demanding protection from the consumer and the consumer demanding satisfaction from the manufacturer by way of insurance claims—and with government involved in all of it.

It is difficult to remember the time—not too long ago—when, if a product did not work, the consumer felt that all he could do was write as scathing a letter as possible to the president of the company and thus at least feel he had gotten "psychological satisfaction." Today's sophisticated consumer will take his satisfaction in cash or check form only, either in a "simple" settlement from the manufacturer's insurance company, or the hard way—through the courts. The court method is especially gruesome since the national news media are also aware of consumer power and eager to give blow-by-blow coverage to corporate court fights with the American homemaker.

The whole area of insurance coverage for design defects is an extremely involved one for the underwriter of insurance. Historically, the insurance industry has maintained that design defects were not insurable. The principle upon which the private insurance sector bases insurability is the *random* occurrence of accidental loss among large numbers of insureds. A defectively designed product or service, by definition, removes the random aspect of injury and loss. And no underwriter would knowingly take that kind of bet. The odds are just not worth it.

On the other hand, insurers have assumed, by way of liability coverage, losses incurred by a manufacturer due to "random negligence" occurring during the production of a product. Primarily due, again, to the development of consumer sophistication and the resultant demands of consumer power, the distinction between design defects and random production negligence has, in the past several years, become a distinction only the minds of the underwriter and the design engineer. The public doesn't really care about "historical concepts of liability." In the minds of consumers and increasingly in the judgment of the courts, the design and manufacture function have merged and are relevant only in terms of a safe, complete new product.

Perhaps the most vivid example of the continuing interrelation between design safety factors and the maintenance of realistic and effective insurance for the consumer is the automobile industry and the highway systems. Design defects in exit and entrance ramps on superhighways continue to cause many accidents, which in turn increases insurance premiums charged the American driver. Auto-repair costs continue to skyrocket, at least partly because designers and engineers had not adequately considered accessibility. On one current model, for example, it is necessary to remove the bumper to replace the taillight bulb, and on another the engine must be lifted out to replace the oil pan. And as repair costs rise so must insurance premiums, and the consumer prepares to attack.

There is a growing body of opinion that holds that no longer must we find individuals liable for automobile accidents, but that these costs must be shared regardless of liability. Increased emphasis is being placed on the need to consider highway safety, auto safety, and the related difficulties of court procedures in handling auto cases as a social problem, with solutions being sought from the federal government. The emphasis is increasingly placed on the overall reduction of accidents, particularly those specifically traceable to engineering and design factors.

The insurance industry has been challenged, particularly in the last two years, to show cause why it should continually be paid more and higher premiums by its customers and to answer for its failure to adequately communicate to the public the underlying causes that have brought about the accidents that give rise to increased premiums. The challenge is justified. Insurance companies have been, to a great extent, unable to suggest viable means of reducing accidents. They have not maintained records that could assist in the better design of either highway systems or safer products. Insurance statistics have been used to determine insurance rates, to attempt to ascertain differences in the accident frequencies between territories, between products, between types of drivers. Perhaps only, and then to some limited extent, in the area of property insurance, where statistics have been kept on very broad types of construction, have statistics been kept in such a manner as to be usable in the prevention of damage or injury.

It is unnecessary to say at this point that we have not done enough to fulfill the public's desire for greater safety in design and engineering. What is more difficult is to say what should be done and where we should go from here.

Where do we go from here? To what degree does the insurance industry make decisions about safety standards? And in what new ways do we need to work with industry, engineers, government, and public-protection and safety associations? Does massive exposure and extension of Underwriters' Laboratories-type inspection and safety "tagging" provide a meaningful way to proceed? Does the caution message on the sides of cigarette packs set a precedent for all consumer products with a health or safety risk involved? And who will determine what degree of risk warrants such a warning? Do we concentrate on developing adequate standards for safety in design in the private sector? Do we wait for government intervention? Do we begin to work more intimately with government to improve safety considerations? And if so, when? Next week? Next year? Who's going to take the initiative? Will intensified scholarship programs in safety engineering provide the answers? Should the public be reeducated to demand safety first—before esthetically pleasing but unsafe design? How ready is the consumer to pay for safety?

All of these questions are asked not to imply that nothing has been or is being done. The insurance industry's concern for safety is well documented. Both individual companies and specialized insurance-sponsored organizations daily add to the list of projects, plans, and programs designed to improve the health and safety of the American public. The questions imply rather that what we have done is not enough. The public is not satisfied with what is being done, and therein lies both the problem and the opportunity for all of us.

TRANSPORTATION OF HAZARDOUS MATERIALS

WILLIAM K. BYRD

In 1968, the Chairman of the Hazardous Materials Regulations Board issued a Notice of Plan to Revise the Department of Transportation (DOT) Hazardous Materials Regulations. Areas for consideration included classification and labels, handling and stowing, placarding and emergency procedures, and packaging. Packaging is the main theme of this presentation. In the Notice, it was stated that packaging requirements should relate to the classification of the material, the quantity involved, and the transportation environment, and that packaging requirements should be stated in terms of performance standards rather than manufacturing specifications. The regulations should also prescribe tests to determine whether the packages meet the requirements.

There are numerous formats and procedures for writing this type of regulation. As a member of the staff of the Office of Hazardous Materials, I am not aware of any DOT-predetermined formats or methods to be recommended. We are still considering the matter in light of comments being received. However, there is one approach that can be discussed in depth. The United Nations Committee of Experts on Transport of Dangerous Goods is currently working on an approach applicable to all modes of transport that may prescribe:

1. General requirements for all packages (applicable to all classes except compressed gases and radioactive materials whether the package is new or is one allowed to be used more than once).

2. Supplemental general requirements for explosives.

3. Particular requirements for explosives.

4. Special requirements for each type of package (e.g., drums, boxes, bags).

5. Testing.

General requirements for packages can include the following:

1. Quality, construction, and adequacy of closures of packages, as pre-pared for shipment, must be such as to prevent leakage caused by changes in temperature, humidity, pressure, or other environmental factors that could exist under normal transport conditions.

2. Packaging used more than once must be emptied and cleaned before refilling.

3. Packagings, new or used, should be able to withstand the prescribed tests successfully.

4. Packaging, including closure devices, in contact with hazardous materials should be resistant to any chemical or other destructive action. The packaging materials must not contain substances that would react with the contents, thereby presenting a hazard, or that would form hazardous products or weaken the packaging.

Plastic materials must not be used if they are likely to be softened, rendered brittle, or otherwise adversely affected by extreme temperatures or chemical action of the material to be transported. The closure and body of plastic packagings must be constructed to adequately resist effects of temperature and vibration occurring during transport.

5. When filling packagings with liquids, sufficient outage must be provided to ensure that neither leakage nor permanent distortion of the packaging will occur as a result of expansion of the liquid caused by temperatures likely to be experienced during transport. In the case of air transport, packagings filled to the appropriate outage should be able to withstand a minimum internal gauge pressure of 15 lb/sq in. without leakage.

6. Packages should not contain different substances that might combine or react dangerously with each other.

7. Inner packages should be packed in such a manner as to prevent their breakage and leakage.

8. Breakable or easily puncturable inner packages should be suitably cushioned and supported in strong outside containers. When a package contains extremely dangerous liquids, the cushioning should include enough absorbent material to prevent any leakage from the package.

In addition to general requirements for all packages, it may be necessary to provide supplemental general requirements for explosives:

1. Handles or handling devices must be provided for heavier packages.

2. Nails or other closing devices that would extend into the interior of the package are prohibited.

3. Outer packages must be such that friction during transport does not generate any heating likely to alter chemical stability of the contents.

4. Closure devices must ensure double protection against leakage.

5. Closures must be such that packages containing wetted or diluted material will not lose the required percentage of water, solvent, or phlegmatizer.

6. Packing must be such that no movement can occur in the package during transport.

7. If explosives are packed with other explosives of a different nature, the packaging must be such that an accidental explosion of any portion of the contents will not cause functioning of the other contents.

8. Each package should be marked with the name of the contents and the net and gross weights of the package.

9. When the package is of the double-envelope type and has voids filled with water that may freeze during transport, a sufficient quantity of anti-freeze must be added to prevent the water from freezing.

Special or particular requirements for packaging intended to contain explosives may be added, such as:

1. When using steel drums, steps should be taken to prevent ingress of explosive substances into the recesses of the seams.

2. The closing device must have a suitable gasket.

3. Explosives must be prevented from coming into contact with the screw threads of closure devices.

4. When metal-lined boxes are used, they should be made in such a way that the explosive substances carried cannot get between the lining and the sides or bottom of the box.

Once a set of general and certain specific requirements are set forth, it is necessary to lay down testing criteria to ascertain whether the package can withstand the rigors normally encountered during transport. Testing procedures can be established by taking into account materials used, production methods, type of hazardous material transported (liquid or solid), and degree of hazard of the material.

Tests required are normally based on methods generally applied at present. Therefore, provision for accepted methods of the future that give identical or equivalent results should be acceptable. Otherwise, we find ourselves again regulating by permit, waiver, or exception.

The kinds of packages to be tested to destruction in any set of regulations would normally be those with a capacity of up to 800 or 900 lb. When we go beyond these limitations, other means of ascertaining the performance of packages may be applied. For example, 30,000-lb casks used in the transport of radioactive materials must meet certain drop-test requirements. However, engineering design certification that a cask will meet a test criterion should be acceptable.

As in any type of testing, the applicability and frequency of tests must be established. At the start of production, tests must be made on each type of construction before use would be authorized. Testing must also be repeated following any modification in design or manner of construction or packing. During production intervals, tests must be repeated to ensure that standards are being met. Before reuse, each packaging should be inspected for corrosion and physical damage. Any package showing signs of deterioration likely to impair its ability to withstand the prescribed tests should be rejected.

At this point, we should discuss preparation of packages and packagings for test. Except in cases of certain tests such as the hydraulic or leakage tests, the tests should be carried out on packages as prepared for shipment. This would include inner packagings. The goods to be shipped may be replaced by non-hazardous materials. If they are replaced, the goods used should have the same density, viscosity, size, and weight; and other physical properties should be similar to those of the goods to be transported.

There are proponents of detailed specifications and proponents of performance standards. To me, specifications in the light of these requirements are general standards. In this whole approach, it may not be necessary to have anything more than a statement, using drums for an example, that sheets for body and heads should be of suitable steel and of gauge adequate to drum capacity and to the service it is to perform. Again using steel drums as an example, other things that may be provided are:

1. Whether a drum must be welded or if it may be folded and grooved.
2. Rolling hoop requirements.
3. Reinforced chimes.
4. Prohibition of beading under rolling hoops or spot welding.
5. Lead-coating thickness and bonding requirements.
6. Requirements for gaskets (taper thread may provide adequate closure).
7. Use of screw-thread closures.
8. Opening diameters for nonremovable-head drums.
9. Sealing devices.

Another bone of contention is the maximum capacity and weight authorized. With modern handling devices, metal drums could very easily be made larger and could handle heavier weights in the future. The advent of the portable bin is one example. The ever-present portable tank is another.

Once a person has reached this point in his thinking, he must look to testing to verify the validity of his approach. There are several types of tests. Some of them are good for one package, and some are appropriate to others. A metal container normally would have to undergo a series of drop tests, a leakage test, a hydraulic test, and a stacking or compression test. Let us follow a series of tests that might be considered appropriate for metal drums. In fact, this particular series is currently being considered for international regulation.

Drop test

NUMBER OF SAMPLES

Three samples to be tested.

PREPARATION OF PACKAGES FOR TEST

Drums intended to contain liquids should be filled with water to 98 percent capacity.

TARGET

The target should be a rigid, smooth, flat, and horizontal surface.

HEIGHT OF DROPS

Solid substances: 4 ft.
Liquids with a specific gravity not exceeding 1.2: 4 ft.
Liquids with a specific gravity exceeding 1.2: 6 ft.

POINT OF IMPACT

The test should consist of two drops:
First drop   The drum should strike the target diagonally on the chime or, if the receptacle has no chime, on a circumferential seam.
Second drop (using the same sample)   The drum should strike the target on a part considered weaker than the chime or, if there is no chime, on a part considered weaker than the circumferential seam. (Closing devices or other parts projecting beyond the chime [or rolling hoops] should also be capable of withstanding the test.)
The testing cycle should be repeated on the remaining two drums.

CRITERIA FOR PASSING THE TEST SUCCESSFULLY

There should be neither leakage from, nor rupture of, the tested drums or the inner packages they may contain.

Leakage test

NUMBER OF SAMPLES

Before every use in transport, each receptacle destined to contain liquids should be tested.

PREPARATION OF PACKAGES FOR THE TEST

No provisions.

TESTING METHOD

The drum should be immersed in water or seams covered with a soap or oil. The manner of maintaining the receptacle under water should not affect the validity of the test.

AIR PRESSURE TO BE APPLIED

The pressure applied should depend on the nature of the material and method of construction used; in any event it should not be less than 3 psi.

CRITERIA FOR PASSING THE TEST SUCCESSFULLY

No air leakage.

Hydraulic test

NUMBER OF SAMPLES

Three samples.

PREPARATION OF PACKAGES FOR THE TEST

No provisions.

METHOD OF TESTING AND PRESSURES APPLIED

Drums should be subjected for a period of 5 minutes to a hydraulic gauge pressure not less than the total pressure (vapor pressure plus partial pressure of inerts present, if any) that could be developed by the contents at the highest temperature likely to be attained during transport. The gauge pressure to be applied should be that of the total pressure likely to be developed during transport multiplied by a safety factor of 1.5.

CRITERIA FOR PASSING THE TEST SUCCESSFULLY

No leakage.

Stacking test

NUMBER OF SAMPLES

Three samples.

PREPARATION OF PACKAGES FOR THE TEST

No specific provisions.

METHOD OF TESTING

Packages must be capable of withstanding for a period of 24 hours a super-imposed weight placed on a flat surface resting on the top of the package and equivalent to the total weight of identical filled packages that could be stacked on it during the transport operations. (A 26-ft stacking test is considered adequate for intermodal shipments since seaborne transport must be taken into account.)

CRITERIA FOR PASSING THE TEST SUCCESSFULLY

No deformation likely to reduce the strength or tightness of the receptacle.

Evidence has been presented that high-frequency vibration may cause cracking in the bottom heads of metal drums. This must be taken into account in future regulation of metal drums. I understand that this type of testing will be suggested for international regulation.

I have heard many reasons given by individuals for support of manufacturing specifications and opposing reasons from those who support performance standards instead of specifications. Those who support manufacturing specifications usually do so because of possible liability involved if they are not told exactly how to build a package, even to the way you hammer a nail or bend a piece of steel. Others feel that with manufacturing specifications, everyone must incur the same expense, and thus the competition cannot lower prices by developing a new technique. However, those who support the performance-standard approach feel strongly that government should set a standard and let industry prove out new designs and methods of construction that will meet the standard. They also feel that new techniques to improve on safety can be developed without the need for continual changes in regulations. This last statement has great merit. At the present time a new innovation in packaging, even one that may provide increased safety, cannot in many cases be implemented without a request for regulation change.

IS TRAFFIC SAFETY IN THE PUBLIC INTEREST?
WHEN? HOW MUCH?

MARTIN WOHL*

Today, engineers and designers increasingly seem to argue that "transport systems must be designed to operate in the safest possible fashion," or that "all actions that will improve transport safety are necessarily in the public interest," or that "something must be done about the safety problem." These views are no less horrifying to me as an engineer than one that attests that all structures, whether buildings, dams, or bridges, must be designed in such a manner that during their economic and functional lives no physical failure endangering the lives of the users should occur. In essence, views of this sort imply that improper engineering design can be defined only in terms of physical collapse and the resulting loss of lives and limbs; also, they imply that any physical collapse is an engineering design failure. Contrarily, I would argue that the absence of more frequent collapse of our civil engineering works can and in some cases will represent bad engineering design of no less importance. (Clearly, though, this latter type of engineering design failure ordinarily escapes attention because the *possibility* of an error in design is not so apparent.) Engineering design failures can and should be defined to include failures of overdesign as well as those of underdesign. Moreover, if an engineer has designed in the face of risk and uncertainty and has properly accounted for the extra gain and costs attendant upon more safety (i.e., with lowering the probabilities of "failure" or collapse), then a physical collapse or collision that occurs purely by chance rather than by oversight should in no way be termed "an engineering design failure."

A similar though slightly different set of attitudes has developed with respect to transport accidents. Particular concern has focused on the "traffic-safety problem" and on the "highway carnage" that accounts for some 50,000 lives and almost 2 million disabling injuries a year. The most

*The research leading to this paper was conducted at the RAND Corporation and under contract with the National Highway Safety Bureau. The opinions, findings, and conclusions expressed in this publication are those of the author and not necessarily those of the RAND Corporation or the National Highway Safety Bureau.

common reaction to these totals is to adopt the stance that "we *must* improve highway and traffic safety" or that "the highway carnage *must* be abated." However common this view is, it is not clear that one may assume that reduction of the highway carnage *per se*, aside from all other considerations, is a necessary goal or one that definitely is in the public interest. Moreover, it is hardly sensible to adopt a policy that attests that "lives and limbs" are priceless. No matter how final loss of one's life or vital parts may be, neither can be regarded as priceless. In a more general sense, our willingness to engage in wars and to commit many "lives and limbs" in Vietnam, Korea, and World Wars I and II is a clear expression of this view. To argue the contrary would be to argue, for example, that people would be willing to sacrifice their homes, recreation, food, clothing, traveling—in a word, everything—in order to reduce traffic-safety hazards and improve their chances of survival and non-injury. Some rare individuals in rare circumstances might be willing to make extreme sacrifices to improve their chances of survival, but most people are willing to take risks because they consider the expected benefits at least important enough to outweigh the hazards, discomforts, inconveniences, and expenses of travel. If their lives and limbs were priceless, they simply would forgo the trip.

And, if lives and limbs were priceless and if improved safety were really an unqualified objective of the traveling public, why do travelers at times *knowingly and voluntarily* increase risk of injury? For example, some drivers will often increase speeds to save time, even though they are aware of the increased dangers of more severe injury and death should they have an accident. And some drivers will continue to drive after drinking even though they know their driving abilities are reduced and their accident potential is increased. In both of these cases, it seems perfectly clear that the driver does not expect an accident to occur but that he simultaneously is aware of the extra risk and feels that the extra risk is worth taking because of the increased benefits.

The foregoing remarks indicate, in a general way, the approach to be adopted in the balance of this paper, and they characterize the nature and types of considerations to be dealt with more fully. However, the specific details and examples will deal almost exclusively with the "auto- or traffic-safety problem" rather than with all types of transport safety situations. Narrowing the scope is deemed useful for at least three major reasons: The problem can be approached with more specificity and thus lend more realism to a subject that often is clouded by generalities. Using virtually any measure of accident hazards, and employing even the most conservative projections, auto accidents (and the accompanying injuries and fatalities) will, over the foreseeable future, outweigh those on other types of transport systems by at least an order of magnitude. And, relative to the opportunities in the auto and highway sector, those safety actions that afford the possibility for significant improvement in nonauto transport systems are probably unattractive when placed alongside the commitments requisite for such improvement.

A Restatement of the Traffic-Safety Problem

Simply, the safety problem is to determine:

- The different safety actions or measures that may be adopted.
- The aggregate levels of benefit and cost associated with the adoption of one safety measure or another.
  - Who will benefit (and to what extent) from different actions.
  - Who will pay (and to what extent) for different actions.[1]

Note that emphasis is placed both upon the aggregate benefit-and-cost totals and upon the incidence of these impacts. Such emphasis is made to ensure that, for example, public officials or engineering designers do not purchase or require more safety for either the public at large or a certain segment of that population than is consistent with the value scales of the people in question. This emphasis also requires that an objective rather than a subjective set of values be incorporated in our design or planning decisions about "how much safety is justifiable."

Admittedly, to discuss different safety actions in terms of the aforementioned benefit, cost, and incidence aspects is hardly straightforward or simple. On one hand, there are problems in understanding what factors affect safety hazards and what the cause-and-effect relationships really are; there are others that arise from the sheer complexity of the driver–vehicle–roadway–environment system and from our inability to explain (and in some cases even to describe) the interworkings of so complex a system; and, of course, there are the usual problems attendant upon a risky or uncertain future world. On the other hand, problems arise simply because of the fashion in which different safety measures affect the traveling public. First, certain types of safety improvements (such as highway redesign) tend generally to improve safety for all of the driving public; that is, the accident frequency or severity rate is reduced equally for all drivers and thus it is the *cumulative* impact on the *entire* traveling public that is of interest. Second, other types of improvements tend to improve safety only for those individuals having or making use of that equipment; examples would be head restraints, lap belts, and shoulder harnesses. For this latter class of improvement, its feasibility is determined by an

---

[1] The term "pay for" is intended broadly and is meant to describe all money and non-money commitments that somebody in the public at large must make in order to improve safety.

examination of the impact on *individual* drivers or passengers; the fact that one individual may obtain a large net benefit from the installation and use of head restraints in no way helps to justify installation of head restraints for another driver who suffers or incurs a net loss from their presence.

Another way to make the above distinction is to say that the first class of safety improvement virtually always will enhance the safety of all travelers and thus produce expected benefits for the entire group (measured, or course, as the accumulation of the expected individual benefits to each traveler), while the costs incurred to make that improvement are not attributable to any one segment of the traveling public but are common to the entire traveling population. The second class of improvement, by contrast, can be judged only in terms of the effects on the occupants of an individual vehicle; in this case, the benefits accrue principally (and perhaps almost entirely) to those having or making use of the special safety items, and, importantly, the costs are directly and only attributable to the vehicle's occupants.

The above distinction between group and individual effects makes it necessary to discuss and consider various safety measures and their safety–risk, cost–benefit circumstances in light of the manner in which their impacts are manifested. One way to accomplish this would be to separate safety actions first with respect to group and individual effects and second with respect to pre-crash, crash, and post-crash effects. For the latter, there are those that affect the probability of having an accident; those that affect the probability of accident severity, given a certain type of accident; and those that affect the probability of rehabilitation of an accident victim, given an accident of a certain severity. To illustrate the differences among various safety actions, their types of impact, and their appropriate analysis, a number of particularized examples of safety actions will be discussed in some detail; mainly, though, the discussion will center on actions that affect either accident probabilities or accident severity, with only peripheral attention being devoted to the matter of rehabilitation of accident victims.

The Role and Analysis of Certain Safety Actions and Federal Standards

Neither the efforts of the automotive industry nor the establishment of the National Highway Safety Bureau will necessarily guarantee the adoption of the most worthwhile level or type of safety improvement.[2] Rather, it seems most likely that private industry will under-provide safety equipment (if left

[2]See, for example, A. Carlin, *Vehicle Safety: Why The Market Did Not Encourage It and How It Might Be Made To Do So,* RM-5634-DOT, The RAND Corporation, April 1968.

to the open market) and that the federal traffic safety program will concern itself more with "how to improve traffic safety" than with "whether to improve traffic safety at all," as well as how much and in what way. Further, federal actions will probably result in over-provision of safety. To illustrate these and other related points, it will be helpful to examine the circumstances surrounding three particular safety measures, the installation of seat belts (both lap and shoulder harnesses), side-marker lights for front and rear side panels of autos, and alcohol standards for auto drivers. However, in examining these examples the major concern is not to test their worth but to introduce and analyze safety measures having different characteristics with respect to private versus public concern and with respect to the necessity or desirability of government intervention. Further, the examination will serve to provide insight into which types of safety actions seem most worthy of government attention, which measures involve different types of system effects, and which system effects and elements deserve more attention in our analyses and evaluations.

SEAT BELTS

Some—perhaps many—feel the auto industry *should* install these (and other) safety devices as standard equipment of its own accord and that its failure to do so stems from the absence of proper incentives or motives. However, to infer, as some have,[3] that auto users are "in no position to dictate safer automobile designs" or that the auto industry in no way feels the pinch would be less than accurate in some important respects. Specifically, the consumer can and often does dictate safer designs—and thus industry does feel the pinch—to the extent that the consumer desires more safety and feels that the increased safety is worth the extra expense. In this instance, the consumer dictates his safety "requirements" or "needs" either positively or by rejecting increased safety when it is offered. To conclude that the consumer cannot dictate safer designs merely because the auto industry does not make some safety device a *standard* item of equipment is, in general, not correct. Rather, failure to make safety devices standard rather than optional items of equipment implies that the auto industry feels the consumer will not find the extra safety worth at least as much as the extra cost of providing it.

However, the consumer or auto user in many instances may not know about or properly consider and account for the full potential benefits (or costs) stemming from acceptance, purchase, and use of safety devices or improvements. Particularly, he is prone to forgo consideration of those aftereffects that do not affect him (i.e., externalities—those gains or losses that are

---

[3] See, for example, Ralph Nader, *Unsafe At Any Speed*, Grossman, New York, 1965, p. viii and pp. 112 ff.

caused by, but not felt by or thought to be the concern of, another group) or those he does not perceive, understand, or care about. Under these circumstances, users may not dictate overall safety "needs" properly, and thus auto manufacturers may not respond properly, at least from the point of view of the public at large. Therefore, some action on the part of government agencies may be appropriate to ensure that all internal—and external—benefits and costs (including unperceived internal ones) will be considered in the process of determining proper safety measures and devices. There are other situations in which government action can be deemed necessary, of course. Some of these will be discussed later. Among possible actions are the setting of design or performance standards, subsidies, and self-insurance requirements (premiums being related to safety equipment and its use).

More specifically, we may charge that buyers and users of autos are not properly informed about the extent to which buying—and using—lap- and shoulder-harness assemblies would lessen the possibility of a fatal or severe injury. And we may argue that the cost of buying and using a seat-belt assembly is a small price to pay for a high chance of surviving a serious accident or of avoiding a seriously disabling injury. On the other hand, the fact that less than 50 percent of those now having seat belts do use them suggests that for many travelers the discomfort and inconvenience of using them outweigh the potential gains. Not unlike the situation facing the public with respect to cigarettes, the auto traveling public may simply—and properly—have been playing an expected-value game, all things considered. That is, auto buyers and drivers (in the past) may have been balancing the costs of buying seat belts now and the discomfort and inconvenience of using belts henceforth against the joint consideration of the probability of having an accident of the type in which the use of a seat belt would reduce or ameliorate the losses or injury; the year in which such an accident would occur; and the reduction in the losses (mental, physical, and otherwise) stemming from seat-belt use given that kind of an accident. To provide some insight on the first of these considerations, one may estimate that on a random auto trip an auto occupant has less than one chance in 60,000 of being injured in an auto accident and has less than one chance in 200,000 of receiving worse than a minor injury, assuming no use of seat belts.[4] Given these odds and our expectations about the year of occur-

---

[4]The chances of being in an accident are higher, of course, but property-damage-only accidents are excluded, as seat-belt use will affect none of the costs or benefits stemming therefrom. Also, these probabilities were estimated by combining information on auto-trip length, on average annual travel, and on recent personal-injury and fatality statistics. Data were obtained from *Accident Facts*, National Safety Council, Chicago, 1968; *U.S. Transportation: Resources, Performance and Problems*, NAS-NRC Publ. 841-5, National Academy of Sciences–National Research Council, Washington, D.C., 1960; *1968 Automobile Facts and Figures*, edited by Hendustan Motors Ltd., Kallman; and H. H. Mitchell, *Emergency Medical Care and Traffic Accidents*, RM-5637-DOT, The RAND Corporation, 1968, Tables 9 and 11.

rence of an accident (e.g., an accident next year is not "half as bad" as one this year), it is not so clear that many auto travelers are underutilizing seat belts; some may be, of course, because of higher accident potential or because of placing higher personal values on "loss of life and limb," and so forth.

A second matter of concern with respect to accounting for all benefits and costs relates to the externalities. That is, would the use of seat belts produce benefits (or reduce costs) for other people? For example, does a father, in deciding not to use belts, properly think about the effect that his death or severe injury might have on his relatives, offspring, and wife? And does he think or worry about the fact that his insurance, income, and savings circumstances may force him or his survivors to become wards of the general public should he be killed or severely injured? Although the answers to these sorts of questions are not readily apparent, one may reason that it is possible and perhaps even probable that travelers may consider them, even though there are no clear-cut signals or price mechanisms to ensure that they are considered. One may reason, for example, that drivers do account for at least some and maybe all of these external effects by observing that they voluntarily purchase life insurance (frequently with double-indemnity clauses) and disability insurance in order to provide income should they become disabled or fatally injured.[5] Obviously, though, not all of their insurance, and thus insurance premiums, can be attributed to automobile safety conditions, and thus only weak hypotheses can be drawn.

When considering the relative magnitude of the direct (or internal) and indirect (or external) benefits stemming from seat-belt ownership and use, it would seem that the direct and personal benefits of using seat belts accruing to the user loom much larger than any external benefits. That is, the user would appear to gain the most from avoiding severe injury or death; it is he who must endure the pain and suffering, who must forgo future pleasures or life itself. Although he probably cannot really appreciate these potential gains, it would seem even more difficult for the user to comprehend the mental grief and anguish that generally would befall his wife, relatives, and offspring should he be injured severely or killed; and it is difficult for him to be responsive to the possibility and costs of him or his family becoming a ward of society. There is at present no effective way for these other people to influence his use of seat belts or for these externalities to influence his behavior (aside from their personal persuasion). For instance, a wife has no effective way to

---

[5] Obviously, this argument presumes rationality in the sense that drivers change their insurance policies once they adopt better "safe driving" habits or buy safer automobiles. Better still, perhaps, would be the emergence of automobile self-insurance programs in which the premiums could be adjusted in accordance with the probability of an accident and with the probability of injury and its severity. See the extended discussion of this point in Carlin, *op. cit.*, pp. 18 ff.

get her husband to take her feelings into account, unless one argues (weakly perhaps) either that she can use nagging or "withholding favors" as a means of reflecting her costs of his not using seat belts or that she can sacrifice part of her weekly allowance to buy him seat belts or "bribe" him into using them.

It is worth re-emphasizing that the driver does have powerful and readily apparent incentives for buying and using seat belts (though there are equally obvious disincentives); it is he who faces the most imminent danger of accident and suffering and it is he who most wants to avoid it. It is he who has the most to gain from using belts, accepting, of course, the earlier discussion about the magnitude of externalities. Once the incentives become strong enough to outweigh the disincentives, seat belts will be bought and used, assuming at least a reasonable understanding of the advantages of seat belts and the possibilities of finding them to advantage. And, of equal importance, the auto manufacturer does have every reason to want his customers to buy seat belts so long as the incremental benefits are at least as large as the incremental costs; quite simply, in this situation he can increase his profit by selling more belts! Also, one may safely assume that auto manufacturers do recognize that seat-belt acceptance and use would, in general, maintain a larger driving and traveling public (i.e., fewer travelers would be killed or permanently taken out of the buying and driving population) and therefore would lead to increased auto sales. However, one can argue that the auto industry has tended to ignore these seat-belt profits because they would adversely affect sales of high-profit sports models to a greater extent, thus resulting in a net loss. Such an argument is conceivable, but tenuous. Further, one probably will find that sportscar users adopt and use safety equipment to an even greater degree than do users of standard models.

As a third aspect impinging on the failure of the market to buy and accept more safety improvements, seat belts in particular, it is necessary to ask why the market isn't operating properly—if indeed it is not—and why any intervention is either necessary or desirable. The various opportunities for intervening must be considered.[6] First, to the extent that there are no externalities and that auto travelers fully perceive all aspects of buying and using seat belts, firms will offer seat belts and auto travelers will buy and use them so long as their incremental benefits outweigh the costs of installation and use. The seat-belt industry appears to be large enough and to be devoid of significant scale economies, thus ensuring (largely) that the price and cost of belts will be driven close together. Second, in those instances in which there are externalities (either with respect to the auto driver or to his passengers) or when it is desirable as an equity matter to protect the lives and limbs of those who cannot afford seat belts, some action or intervention might be appropriate. Among the possibilities are:

[6] Again, Carlin, *op. cit.*, covers this subject in more detail.

Component and design standards.[7]

Subsidies to auto owners to induce more seat-belt installation and use.

Use of information or counsel as a means to better inform auto travelers of the effects of belt use.

Self-insurance requirements in which premiums are related to safety equipment and its use.

The first possibility has the disadvantage of causing the installation of some belt assemblies that will seldom or never be used; thus, resources will "needlessly" be wasted. And one can argue that owners having belts installed involuntarily will tend to use them less than will those who installed them voluntarily (though prior nonuse may have stemmed merely from inability to afford them rather than from dislike of their use or failure to understand their advantages). However, one may also argue that simply having the belts installed—even on an involuntary basis—will tend to increase use. Further, by inducing belt assemblies to be installed for all passenger occupants, passengers will be protected when riding with drivers or owners who otherwise would not consider the passenger's well-being.

The use of subsidies might provide a useful mechanism for inducing installation and use, though some perverse income transfers might accompany such a mechanism, depending on the conditions placed on granting the subsidy. On the latter point, seat-belt ownership is highly correlated with occupation, which serves as a proxy for income level.[8] If the subsidy, or at least its extent, were conditioned upon income, it might be possible to avoid perverse income transfers.

The third possibility, that concerning information and counsel for auto owners and travelers, appears to offer some reasonable hope for increasing seat-belt installation, though not necessarily use. For example, the work of Bass and Wilson easily leads one to conclude that "gentle persuasion" on the part of the physician can lead to increased seat-belt installation at all income levels.[9] However, it should be pointed out that no information is available on the extent to which the increased installation affected the extent of use.

The fourth possibility, that of requiring self-insurance in which the premiums are related to the safety equipment and its use, offers obvious advan-

[7]Although the present federal standards with respect to belt assemblies are termed "performance standards," they should be looked upon more appropriately as design standards. As such, they probably tend to inhibit privately sponsored research on the development of other types of restraint devices.

[8]See L. W. Bass and T. R. Wilson, "The Role of a Physician's Influence on Installation of Seat Belts," in *The Seventh Stapp Car Crash Conference Proceedings*, Charles C Thomas, Springfield, Illinois, 1965.

[9]*Ibid.*

tages for internalizing the external effects stemming from disablement and death.[10] This possibility would provide financial incentives for drivers to install and use belt assemblies and to balance out the pros and cons according to their own personal circumstances. The major difficulty, of course, still would rest with the establishment of the "appropriate" level of self-insurance to be required—particularly in view of the fact that such a judgment rests squarely on the valuation of "life or limb" losses and of the pain, grief, and suffering.

SIDE-MARKER LIGHTS

A significant number of accidents, both fatal and nonfatal, involve two motor vehicles entering intersections at an angle (that is, from intersecting or crossing roadways). This group is the most significant one for urban situations and about the second most significant one for rural cases. Although data are not available to segregate the two-car intersecting-angle accidents into those occurring at night and those occurring during the day, one can hypothesize that a major proportion of them occur during evening hours. One may also hypothesize that prior to the inclusion of side-marker lights on most vehicles a sizable number of the nighttime accidents will occur because one (or both) of the drivers is not able to see the headlights or outline of the other car. For example, consider the situation at a right-angle intersection at which vehicle A is approaching along the main roadway from the west and vehicle B is attempting to enter the main roadway by turning right from the south. In this case, it may be difficult for vehicle A to see vehicle B during evening hours (assuming for the moment that the streets are not well lighted or that vehicle B is not equipped with side-marker lights). Moreover, if vehicle B can easily see the headlights of vehicle A, he may, through overconfidence, be led to believe that vehicle A can also easily see him; this may cause him to enter the intersection too soon and be hit. These two-car intersecting-angle accidents are not the only ones in which difficulties in seeing the other vehicle arise; for example, a similar type of situation would present itself for nighttime accidents involving nonintersection two-car crashes between a moving vehicle and one pulling out of certain parking spots or one entering from an alley.

For situations such as those outlined above, it is appropriate to consider, first, the usefulness of side-marker lights mounted on the front and rear side panels of autos and, second, the private-market and governmental forces affecting their installation. Without much question, the installation of side-marker lights—similar, say, to those required by the 1968 federal vehicle

---

[10]Carlin, *op. cit.*, argues the validity of this type of insurance in connection with his discussion of the Keeton-O'Connell plan, pp. 19 ff.

standards and now appearing on autos produced after January 1968—will enable some of the above-described accidents to be eliminated and some to be reduced in severity, particularly as a major segment of the automobile fleet is equipped with these devices. If we assume that neither the amount nor the character of tripmaking will be affected by the presence of this sort of safety device, then the feasibility and desirability of their installation can be judged in the usual fashion—by examining the gains from saving "life, limb, and property," as well as those from reducing grief and suffering, and subtracting the incremental costs of side-marker-light installation. However, in examining the feasibility question and the forces governing its installation, it is seen that this type of safety device differs distinctly from the seat-belt example. First, there are clear and significant externalities in the side-marker-light case, both from the point of view of the car owner and user and from that of the automotive manufacturer. Second, arguments in favor of requiring federal component standards for side-marker lights appear considerably stronger than those for seat-belt assemblies, though other alternatives for inducing widespread installation of side-marker lights should also be considered (e.g., subsidies).

With respect to externalities, the side-marker case differs from the seatbelt case in that side-markers affect both the first and second collision probabilities (i.e., both the probability of an accident and that of injury severity), and they affect these probabilities both for the car having the side-marker lights and for other vehicles that come in contact with that car. Thus, the existence of side-marker lights on one car, for example, will work to the benefit of other cars as well; as a consequence, side-marker lights will bring about not only the direct benefits of riding in a safer, more easily seen vehicle but also indirect benefits to other cars on the road. The latter group of benefits is, of course, external to the car owner buying the side-markers; and, of some importance, one can argue that the external benefits will be of the same magnitude as the internal ones. Under these circumstances it should be evident that under the free-market system, auto owners will tend to understate the benefits stemming from purchase of side-marker lights and will tend to under-buy them—from the point of view of the welfare of the general public. By the same token, the existence of these external benefits would tend to dampen the enthusiasm of the automotive industry and to inhibit an attempt to make side-marker devices standard equipment on automobiles. If car buyers were unwilling (for whatever reasons) to take account of, or consider, other than their own direct personal (i.e., internal) benefits in assessing the worth of side-marker lights, then the auto manufacturer, by making them standard equipment, would stand to lose only if buyers (on the average) were willing to pay an additional price of less than the incremental cost.

Turning to the second point, let us assume that the total external and internal benefits stemming from installation of side-marker lights on all autos outweigh the incremental costs of their installation, but that the internal benefits by themselves are less than the incremental costs. The absence of side-marker lights as standard or optional equipment prior to the 1968 federal vehicle standards suggests that the assumption is valid. One should then consider the forces governing their manufacture and purchase as well as the institutional mechanisms that might be used to overcome market imperfections. First, the existence of external benefits stemming from installation of side-markers on any auto, which we may assume are approximately as large as the direct internal benefits, will lead to under-utilization of these devices if the free market is relied upon for their installation, unless, of course, the auto industry or others cross-subsidize this safety feature. Second, there would appear to be significant economies in having side-marker lights installed during the manufacturing phase rather than having them added by repair or body shops after purchase; this type of economy for side-marker lights would appear to be significantly greater than that with respect to seat belts.[11] Third, and of perhaps no little importance, the use of side-marker lights as standard equipment (particularly when virtually all autos are so equipped) would in all likelihood reduce the overall sales and profits of the automotive industry to a measurable and perhaps significant extent. That is, if, as hypothesized, the use of side-markers did lead to the prevention of a sizable number of two-car "right-angle" intersection accidents, fewer automobiles would be scrapped and replaced with new vehicles. Given this assumption, there would be no economic incentive for the automotive industry to provide side-markers, particularly if all manufacturers so equipped their vehicles. (No single auto manufacturer would dare offer this device to gain a competitive market-splitting edge; if he should, other manufacturers would probably find out about it, and over the long run, auto manufacturers individually and collectively would lose sales and revenues.) However, if one took into account the long-term effects of installing side-marker lights—the reduction in accident and suffering costs, the ultimate reduction in insurance premiums, and the offsetting increase in auto sales that might come about, depending on market elasticities—one may find that over the long run the profits of the automotive industry would remain about the same, with or without side-marker lights. Still, it is difficult not to conclude that the auto industry would suffer both short- and long-term losses as a result of side-marker lights, all things considered; the industry (probably) was at least intuitively aware of this in deciding not to install these devices voluntarily.

Side-marker lights serve as a classic example of a situation involving externalities, in which the free market cannot (necessarily) be expected to operate in the best interests of the public at large. Some mechanism is needed that

[11]Carlin, *op. cit.*, makes this point well in his *Memorandum* (RM-5634-DOT); the contrast he notes is even more marked for improvements such as energy-absorbing steering columns.

would confront the auto buyer and the auto industry with the external and in-
internal benefits and costs stemming from the installation of safety equipment
and would ensure that buyers do indeed consider both. In addition, of course,
for those individuals who may value their own internal benefits on a low basis
and who either do not find the external benefits large enough to justify their
purchase or do not consider them, society may desire for simple equity reasons to force them to add side-marker lights, simply to protect the lives, limbs,
and property of "the innocent." Among the possible mechanisms for properly
introducing the externalities, as well as internal consequences, are:

Public information programs.
Public subsidies for side-marker lights (at least) to the extent of any deficit
between incremental costs and internal benefits.
Federal component standards requiring all autos to be equipped with side-
marker lights.

The practical effect of mounting a broad-scale information program, for
instance to inform travelers of the risks of injury or death with and without
using side-marker lights, can hardly be gauged with any assurance. The Bass
and Wilson study does at least suggest the potential of a direct appeal program, but the poor showing of Ford sales over a decade ago despite the addition of safety features and a substantial advertising program warrants some
caution with respect to the effectiveness of such campaigns.[12]
If we assume tentatively that information programs should be used in conjunction with one of the other two mechanisms mentioned rather than relied
upon exclusively, we can scarcely choose between the second and third possibilities. Either may lead to all or virtually all autos being provided with
side-markers, to almost the same amount of car sales, and to about the same
level of public-wide benefits and costs. This result implies that auto sales are
inelastic with respect to a price rise equivalent to the difference between incremental costs and internal benefits. However, a more probable view is that
mandatory installation of side-marker lights with no public subsidy will cause
auto manufacturers to experience a drop in overall profits (which requires us
to assume that buyers are unwilling to pay more than the internal benefits of
side-marker lights, that the incremental costs of their installation are greater
than the internal benefits, and to assume also that manufacturers will not
raise the price of other car parts to cover this deficit); or, as indicated, it may
cause them to balance out the deficit by hiking the price of other car parts to
cross-subsidize the deficit from side-marker lights. In either case, of course,

[12] For a discussion of the latter point, see the section of Carlin's *Memorandum, op. cit.*,
entitled "Safety Doesn't Sell."

the auto manufacturer will have to cross-subsidize the deficit, one leading to maintaining roughly the previous level of profit and the other causing a drop in profit. Should the auto industry be forced to take less profit in the process of providing side-marker lights (as required by the federal standard), an income transfer is implied from auto-industry workers and stockholders to auto travelers; that is, the cost of the devices would exceed the payments by auto buyers (payments we will assume are equal to their internal benefits), the deficit being borne by the auto companies. Under these circumstances, it is hardly clear why it is equitable to undertake actions that may force income transfers of this sort. In fact, since the external benefits accrue not only to auto owners but to all auto travelers (drivers *and* passengers, whether of the same family or not) and thus are widely spread throughout the economy, it might seem more reasonable to use federal subsidies to underwrite the deficit, perhaps through increased user gas taxes. Obviously, this is to argue that this class of safety device can be looked upon as a public good of sorts.

ALCOHOL STANDARDS FOR AUTO DRIVERS

Numerous studies have addressed the "drinking-driver" problem in an attempt to establish the effects of drinking on driving capability and accident potential.[13] However, these studies must be regarded as less than conclusive for a variety of reasons. Numerous analyses have documented the extent to which fatally injured drivers were drinking, the figures generally ranging upward from 50 percent for all fatal accidents and from 70 percent for single-vehicle fatal accidents. Moreover, analyses have shown that the preponderance of these accidents involve drivers with very high alcohol concentrations; for instance, Haddon and Bradess fround that for one-car fatal crashes in Westchester County about 50 percent of the drivers had blood alcohol levels of 0.15 percent or more (which, on the average, would be equivalent to about ten one-ounce drinks of 80-proof liquor) and that less than 5 percent had blood alcohol levels below 0.05 percent (about five one-ounce drinks).[14] Numerous other studies (which for the most part deal with urban situations) have found that drivers deemed responsible for fatal accidents have even higher blood alcohol concentrations.[15] On the other hand, one can be less than sanguine

[13] For a careful documentation of the relevant statistics, see H. H. Mitchell, *Alcohol and Traffic Accidents*, RM-5635-DOT, The RAND Corporation, 1968.

[14] W. Haddon, Jr., and V. A. Bradess, "Alcohol in the Single Vehicle Fatal Accident: Experience of Westchester County, New York," *JAMA* 169:1587–1593, 1959. In many states a driver with a blood alcohol level of 0.15 percent is deemed to be "drunk."

[15] See J. R. McCarroll and W. A. Haddon, Jr., "A Controlled Study of Fatal Automobile Accidents in New York City," *J. Chron. Dis.* 15:811–826, 1962; and M. L. Selzer and S. Weiss, "Alcoholism and Traffic Accidents; Study in Futility," *Amer. J. Psych.*

about the extent to which the analysis supports inferences with respect to cause-and-effect relationships between alcohol consumption and accidents, or with respect to the effectiveness of controlling blood alcohol levels of drivers. This somewhat tentative conclusion relies principally on the following three factors:

The data and studies deal largely with fatal accidents rather than with the more general accident problem.

Data used for control purposes and to validate the effects of drinking on accident potential usually include the nondrinking drivers observed to be traveling at that site and time of day rather than a group of drivers having overall drinking characteristics similar to those of the accident-involved drinking drivers but having low blood alcohol concentrations at that site and time. Essentially, then, most alcohol-accident investigators base their inferences and analyses on different populations (drinkers and nondrinkers, which by definition are different) rather than on characteristics of the same population under the two conditions of driving while drinking and driving while not drinking.

Evidence increasingly leads one to conclude that from one-third to three-quarters of the drinking drivers having accidents are "problem drinkers."[16] Moreover, the data and studies are incomplete with respect to urban versus rural conditions and, in fact (as noted), deal largely with the urban situation.

As for the first of these three points, it should be clear that a broader and more comprehensive analysis is necessary, both for a proper understanding of the drinking-driver problem and for consideration of practical countermeasures. Only one-third of 1 percent of the traffic accidents involve fatalities—and, in turn, only some fraction of these involve heavily drinking drivers, whether responsible or not—and less than 3 percent of those persons injured in accidents are killed. Moreover, about 70 percent of the traffic deaths occur in rural areas (the area in which knowledge is weakest and in which driver control and enforcement most difficult and expensive). Taking all these into account, how effective will various alcohol-control and enforcement programs

---

122:762–767, 1966. However, it would seem wise to view data relating blood alcohol levels to accident responsibility with a critical eye because of the obvious "circularity" or *a priori* dependence of the latter on the former; that is, it is difficult not to suspect that traffic officers or others judging responsibility or filling out accident reports are biased against the drinking driver, aside from other evidence.

[16] See, for example, McCarroll and Haddon, *op. cit.*; Selzer and Weiss, *op. cit.*; and J. A. Waller and H. W. Turkel, "Alcoholism and Traffic Deaths," *N. Eng. J. Med.* 275:532; for further discussion of this point, see the section of Mitchell's RM-5635-DOT, *op. cit.*, entitled "Degree of Involvement of Problem Drinkers."

be in terms of economic efficiency and of significantly affecting the accident totals? It is clear that controlling and enforcing the driving habits on 2.8 million miles of surfaced road in this country (83 percent of which is rural) is considerably more complicated than in Sweden, which has only one-fortieth as much road mileage as the United States, or in the United Kingdom, which has only one-twentieth as much. By comparison, all of Europe has only 1.4 million miles of paved road.

The point relating to analyzing control data and populations presents a significant stumbling block in the drawing of valid statistical inferences, particularly when analyzed in light of the point about problem drinkers. The ideal control group would consist of the identical group of accident-involved drivers but who have low blood alcohol concentrations. This observation seems particularly important when evaluated alongside the findings of Allsop and Borkenstein, as summarized by Mitchell.[17] These two studies indicate that moderate drinkers, when driving while their alcohol levels were low, had lower accident-involvement indices than did either lighter or heavier drinkers, and that in some situations driving after a moderate amount of drinking did not produce more hazardous accident potential.[18]

The point concerning problem drinkers appears to be of major significance in any analysis and evaluation of the effects of alcohol on accident potential. The studies of McCarroll and Haddon, *op. cit.*, Selzer and Weiss, *op. cit.*, along with others by Bjerver *et al.*, by Popham, and by Schmidt and Smart,[19] document with a fair degree of conclusiveness that problem drinkers dominate the accident-involved drinking-driver group and that these problem drinkers have, according to Selzer and Weiss (based on their small-sample Ann Arbor study),

had a long history of serious psychopathology which may well have contributed to their accident susceptibility. They were frequently paranoid (52 percent), violent (28 percent), depressed (28 percent) or suicidal (14 percent).[20]

---

[17]"Accident Involvement of Drinkers When Not Drinking," in RM-5635-DOT, *op. cit.*

[18]For other comments on differences between light, moderate, and heavy drinkers as they relate to certain physiological characteristics, see W. Haddon, Jr., "Alcohol and Highway Accidents," in *Proceedings of the Third International Conference on Alcohol and Road Traffic*, September 1962, British Medical Association, London, England, 1963.

[19]K. B. Bjerver *et al.*, "Blood Alcohol Levels in Hospitalized Victims of Traffic Accidents," pp. 92–102, in *Proceedings of the Second International Conference on Alcohol and Road Traffic*, Alcoholism Research Foundation, Toronto, Canada, 1955; R. E. Popham, "Alcoholism and Traffic Accidents," *Quart. J. Stud. Alcohol* 17:225–232, 1956; W. S. Schmidt and R. G. Smart, "Alcoholics, Drinking and Traffic Accidents," *Quart. J. Stud. Alcohol* 20:631–644, 1959.

[20]Selzer and Weiss, *op. cit.*

The above findings raise some obvious but important questions for serious consideration: Would problem drinkers not under the influence of alcohol exhibit better or worse accident-potential tendencies? Would problem drinkers, if forced to restrict their drinking so that they were never under the influence of alcohol while driving, develop even worse psychopathological tendencies and perhaps inflict them on other people in other situations (to the point of bodily or psychological harm), thus offsetting the gains from reduced traffic accidents?[21] Answers to these questions have yet to be subjected to intensive analysis. Schmidt concluded that "the difference in accident involvement between alcoholics and the general driving population would seem to be solely a function of driving behavior after the consumption of alcohol."[22] Mitchell voices a more cautious attitude in saying, "*If* it can indeed be shown that alcoholics and heavy drinkers, when sober, are not an increased accident risk, then countermeasures against the *driver when drunk* may be feasible."[23]

Turning to other important aspects of controlling drinking drivers, there seems to be little question that alcohol can and does impair visual acuity, reasoning, and other physical capabilities, and does cause some accidents. But there also is no doubt that driving ability and accident potential can be, and are, affected or impaired by other factors such as age, lack of experience, inattention, fatigue, and so forth. The more general problem—and perhaps the more equitable attitude to adopt—is to develop techniques for detecting and controlling the impairment of driving capability, particularly that impairment that reaches "dangerous levels" or results in drivers imposing costs on society that are larger than their private payments, *whether that impairment is caused by drinking, old age, lack of experience, or any other factor.* The matter of establishing the cutoff point, or the point at which driving behavior is dangerous, is no mean task and is, of course, central to this discussion. We should focus our attention on determining the relationship between different levels of driving ability and impairment as they affect traffic safety and as they can be detected and controlled and on determining the costs, payments, and benefits associated with the different levels and detection–enforcement mechanisms.

---

[21] Recently it was reported that Britain's new stringent "drunk driving" law, while (perhaps) reducing road deaths by 20 to 35 percent, has been blamed for an increase in sex crimes. "Citing records showing 300% increase in sex crimes, the [Birmingham police surgeon] said emotionally disturbed persons drank alcohol to control their inhibitions but the breathalyzer law has reduced the drinking habits and hence the control." *Amer. Med. News* 11:1, 1968.

[22] V. Schmidt *et al.*, "The Alcoholic Driver," in *Proceedings of the Third International Conference on Alcohol and Road Traffic*, September 1962, British Medical Association, London, England, 1963.

[23] H. H. Mitchell, RM-5637-DOT, *op. cit.,* p. 21. (Emphasis added.)

Putting the matter in more blunt terms, some segments of our society and many accident researchers seem more concerned about legislating against the drinking driver as a moral proposition than about detecting and controlling dangerous drivers, whoever they may be and for whatever reasons their driving behavior is poor.

In general, then, it appears reasonable to argue for more attention to the identification and measurement of overall driving capabilities—to mechanisms for measuring driver response, driver reactions, visual acuity, and so forth—without regard for the factors that cause inability to drive; certainly, this view would contrast with the simpler proposition of detecting and measuring blood alcohol concentrations. This broader view seems to be validated, even with respect to blood alcohol concentration tests, because of wide variations from person to person stemming from basic differences in body weight, metabolism, and time since alcohol ingestion, as well as in alcohol consumption.[24]

Of equal if not greater significance in judging the wisdom of controlling drinking drivers and imposing rigid standards with respect to blood alcohol levels, it should be (but seldom is) recognized that the imposition of more stringent alcohol standards will probably result in reducing benefits as well as accidents and fatalities (assuming that increased alcohol consumption by itself causes more accidents).

If more stringent blood alcohol standards (say, reducing the "drunk" level from 0.15 percent to 0.05 percent) were imposed, some of the former heavy-drinking drivers would continue to drive as before but would drink less and stay within the law. For these people, and those who would have been adversely affected by their actions, the accident costs (broadly construed) would be reduced *if* the driving abilities of the drinkers were enhanced by having a lower alcohol content. (This condition must be proved, however.) In addition, their long-run medical costs would probably be reduced (fewer physiological, and maybe psychological, problems). But for most of these drivers, benefits would be reduced by inhibiting alcohol consumption. One must, for example, assume that they will gain less pleasure from drinking less. Another group affected would be those who drove and drank heavily before the imposition of more rigid alcohol standards but now decide not to make the trip at all rather than inhibit their drinking. For these people, a reduction in net benefit would probably result. That is, before the higher standards are imposed, they derive net pleasure from being unrestricted when traveling, drinking, and enjoying their day-to-day activities; otherwise, they would not behave in this fashion. Note that rationality on the part of the drinking driver—in terms of his private or internal benefits and costs—is assumed; for all but the alcoholic driver, this

---

[24] A summary of various aspects of this problem appears in the section of Mitchell's RM-5635-DOT, *op. cit.*, entitled, "Amount of Drinking and Blood Alcohol Levels."

assumption seems valid. For the alcoholic driver, though, some caution must be expressed with respect to the assumption. Afterwards, their previous net pleasures would be forfeited, since these people would presumably decide that the reduction in overall value or benefit from making the trip—a reduction that stems from having their pleasure from drinking reduced—is so large as to make the entire effort, time, and expense of traveling less than worthwhile. For them, of course, there would be a net loss resulting from the change in legal alcohol content. Offsetting these losses, though, would be gains stemming from fewer accidents for others. Again, the balance is not self-evident; neither is it evident for the entire group of drinking and nondrinking drivers involved. In short, the imposition of more rigid controls over driving under the influence of alcohol would not necessarily represent an action in the public interest.

Two final points are relevant. First, some would argue that the elderly or physically impaired driver does not have the same choice as the drinking driver, the latter being able to improve his driving capabilities merely by curtailing drinking and thus having at least a qualitatively different personal responsibility for an accident. However, from the standpoint of the occupants (and those caring about them) of other vehicles involved in accidents caused by impaired drivers, there would seem to be little difference between being killed or maimed by a drunk driver and being killed or maimed by a driver with some other physical disability. Second, to the extent that these two groups can be identified and controlled with equal ease (and cost) and to the extent that the reduction in accidents is the same for controlling or restricting their driving, the remaining argument for distinguishing between them, other than strictly a moral one, rests upon differences between their losses or reductions in travel benefits resulting from less travel, from travel by a different mode, or from traveling with restrictions on drinking. Of course, one may hypothesize that elderly or physically impaired drivers would suffer more than would drinking drivers and thus that different driving standards should be established and enforced for the two groups; obviously, analysis should be undertaken to test such a hypothesis.

SUMMARY OF ESSENTIALS OF THREE EXAMPLES

For the seat-belt-assembly case, one can argue that overt action is required only to the extent that car owners and travelers are not well informed about, or understanding of, the injury- and death-reduction potential of such devices and that they do not fully take account of gains for themselves and those of persons caring about them with respect to such things as reduced pain, grief, and suffering. Failure to account for these matters results in externalities, suggesting that government action of some sort is desirable so long as the total external and internal benefits outweigh the additional costs. However, if one contends that drivers carry life insurance and disability insurance in recognition of these externalities, that such a practice is widespread, and that knowledge about the advantages of seat-belt assemblies is common, there would be little justification or necessity for government intervention or for making seat belts a standard item of equipment.

For the side-marker-lights case, all of the above arguments are pertinent and should be considered. But, in addition, we must consider the fact that the occupants of other vehicles will be affected by the installation of side-marker lights on our vehicles; under present circumstances, the benefits that would thus accrue to the occupants of the other vehicles (and to any other parties having an interest in their life and health) would be external to and thus would be unaccounted for by car buyers and auto manufacturers. Further, to the extent that the direct internal benefits stemming from purchasing side-marker lights are less than the costs of their installation, and as long as the sum of external and internal benefits is at least equal to the incremental costs, the public interest would be served by inducing or enforcing their installation. Under these presumptions, a strong case can be made for use of federal component standards or subsidies (the latter being financed through user gas taxes, for instance) to bring about their widespread use.

For the alcohol drunk-driving standard case, an even more complex set of issues is involved. First, externalities (or at least potential ones) of the same kind outlined for the seat-belt and side-marker-light cases arise when considering the effects of changing drunk-driving—or, more generally, dangerous-driving—standards. Second, and in contrast to the previous cases, an altering of standards can and probably will affect the amount of, and the pleasure or utility derived from, tripmaking. As a consequence, a reduction in travel benefits, which would inevitably result from the imposition of more restrictive drunk-driving or driving-capability standards—in addition to safety considerations—must also be included in the overall system-analysis and evaluation calculus. Aside from moral judgments, it would appear that the strength of the case for government intervention would lie somewhere between that of the seat-belt-assembly case and that of the side-marker-light case.

Other Aspects of the Analysis and Evaluation of Safety Measures

In the previous sections, the primary objective has been to outline and describe some of the more important factors of the analysis and evaluation of safety measures. The major message has been to emphasize that safety *per se* should not be viewed in isolation, that neither designers nor society should view safety as a necessary good regardless of all else, that governmental intervention or control is not always necessary and in some instances may tend to undermine the general welfare, that the mere lack of safety progress on the part of the automotive industry is not necessarily a sign of not acting in the public's interest, and that equity issues surround the traffic safety problem.

It now seems useful to discuss, in a more technical fashion, aspects of benefit–cost or evaluation procedures and matters of efficiency as they relate to safety measures.

In any assessment of particular safety measures, it is important to differentiate between measures that affect only accident-related costs and measures that affect both costs and benefits. Some safety measures may affect only the costs associated with traveling and accidents, either because drivers and auto passengers perceive no change in their private costs or payments or because they do not value the trip any more highly due to the extra safety.

We must distinguish between safety measures that have virtually no effect on the amount and character of tripmaking (i.e., the volume of trips, their origin–destination pattern, time of travel, and so forth) and measures that do affect the amount and character of travel (i.e., more trips are made because of the reduction in accident hazards, and so on).

In virtually all benefit–cost and evaluation studies the analysis is restricted solely to an examination of incremental cost increases or reductions that stem from introduction of safety measures; the analyst generally regards incremental cost reductions as incremental benefits and virtually always assumes that the safety measure will not affect the amount of trips or their character. Such an analysis is probably incorrect and certainly incomplete. To regard incremental cost reductions as incremental benefits falls far short of a strict interpretation of the change in benefits. One must, in addition, look at the increment in value or in "willingness to pay" (measured in whatever terms) on the part of the traveler and other interested parties. A reduction in property damage, or in medical facilities and services required, to take another example, is of course a cost reduction, which can be balanced against any cost increases from safety improvement. On the other hand, the increment in value resulting from a reduction in pain, grief, suffering, and loss of life or limbs is a legitimate benefit; but this incremental benefit should not be assessed by a simpleminded evaluation of the discounted earnings that otherwise would have been forgone, the practice most often adopted. Perhaps the most concise and appropriate statement regarding the use of incremental discounted earnings saved as a proxy for incremental benefits stemming from safety measures is the following:

There is no reason to suppose that what a man would pay to eliminate some specific probability, P, of his own death is more than, less than, or equal to, P times his discounted expected earnings. In fact, there is no reason to suppose that a man's future earnings, discounted in any pertinent fashion, bear any particular relation to what he would pay to reduce some likelihood of his own death.[25]

To accept this judgment is to reject the principal benefit measures now used in safety-oriented benefit–cost studies and to necessitate the adoption of an entirely new approach for evaluation of safety measures.[26] In brief, it is to require, in addition to knowledge about the cost consequences, an understanding of transportation demand (that is, the time-dependent relationship between quantity of tripmaking and the entire range of socioeconomic variables, e.g., travel price, travel service, and so on) and of interactions between transportation demand and the physical, economic, and social environment.

Although the relationship between safety measures and the amount of tripmaking is commonly, if not always, ignored, it represents a difficult aspect that must be introduced into safety analysis and evaluation studies. Some safety or joint safety–travel-service measures, such as alcohol (drunk-driving) standards or major highway-design improvements, probably will substantially change both the amount of tripmaking and the travel benefit totals. As a consequence, to assess only the accident reductions—and cost increases or reductions resulting therefrom—while assuming that the amount of tripmaking remains constant is misleading. To compare the amount of congestion and accidents that would have resulted without, for instance, the Interstate Highway System with the amount that resulted with it—while assuming the same amount of tripmaking—would be an exercise in futility.

The difficulties in forecasting travel changes that will result from making safety or joint safety–travel-service improvements in no way minimize the necessity for at least attempting to estimate those changes. A study without them will certainly be incomplete and may offer benefit–cost ratios or net-present-value totals that will lead to incorrect decisions, either in terms of whether to do anything or in terms of selecting the best program or project.

With respect to gaining knowledge about transportation demand, about travel forecasting, and about the value of avoiding pain, grief, suffering, and loss of lives and limbs, it should be pointed out that the preservation of product differentiation (and clear information on the differences) and of apparent pricing mechanisms is of major significance. That is, to have a well-structured and competitive marketplace with a variety of safety packages, well differen-

[25] T. C. Schelling, "The Life you Save May Be Your Own," in *Problems in Public Expenditure Analysis*, Brookings Institution, Washington, D.C., 1968. This statement applies only to deaths, but similar reasoning would apply to injury situations.
[26] Two pertinent examples of studies so rejected would be "Motor Vehicle Injury Prevention Program," U.S. Department of Health, Education, and Welfare, August 1966; and L. Lave and W. Weber, "A Cost-Effectiveness Analysis of Auto Safety Features," research report prepared under a grant from the National Safety Council, September 1967.

tiated in terms of safety offerings and tailored to consumer preferences, and to have a well-identified price structure are necessary for determining with any degree of specificity the nature and characteristics of consumer demand. In those instances in which externalities distort the market, or where competition is imperfect, governmental action can be taken to strengthen the workings of, and our knowledge about, the market. However, it seems clear that increasing use of nonhierarchical vehicle-design standards, such as those now being proposed and imposed by the National Highway Safety Bureau, will seriously inhibit our abilities to understand the demand behavior and thus to properly analyze and evaluate various safety measures; that is, we are increasingly neglecting or discarding the basic tools and information so necessary to forecasting travel and determining the value of safety to the public.

There can be no perfect marketplace with respect to the pain, grief, suffering, and loss of life or limbs resulting from traffic accidents. The disutility of such experiences cannot be appropriately measured either in advance of the event itself or merely by observing or reflecting upon the plight of others. But the inability to characterize the offerings of the marketplace or to give people the exact information necessary to reflect upon alternatives does not mean that no attempt should be made in that direction.

In addition, more efficient mechanisms might be adopted for informing drivers about the consequences of their specific driving patterns. Our concern should be directed not only at an appropriate evaluation of different safety measures, but also at better informing the driver, as he makes his individual day-to-day driving decisions under varying physical and mental conditions, of the effects of taking different actions, of driving in one condition or another, and so on. Placed in these terms, proper pricing mechanisms can (under the appropriate circumstances) provide measures for evaluating design or program changes and can induce drivers to take into account the ill effects of their actions on others as well as on themselves. For example, drivers may sometimes make trips (let's say after drinking) simply because they think they are more capable of driving carefully than is in fact the case; given better information, they may prefer to alter their actions, or driving behavior, voluntarily.

Some Summary Comments About Different Safety Measures and Programs

The range of potential safety measures, and the possible institutional mechanisms for their introduction, is wide. To review all of these possibilities systematically and to analyze and evaluate their overall effectiveness is clearly a task beyond our present capabilities. It does, however, seem useful to set down those safety measures and programs that seem most worthy of intensive analysis.

These measures and programs fall into four distinct areas:

Highway, vehicle, or control devices (or operations) designed to prevent accidents

Vehicle and highway features designed to ameliorate injury or prevent death, given an accident

Programs designed to salvage or rehabilitate the injured accident victim following an accident

Programs designed to compensate victims or their survivors following accidents.

Thus far, the attention of the research-and-development community has focused primarily on the first two areas, with respect to both vehicle and highway design. Examples include Interstate Highway System development, highway-design changes (such as those pertaining to alignment, lateral clearance, interchange design, median barriers, and surface texture), better braking systems, interior padding and redesign, belt assemblies, and energy-absorbing steering columns. To focus our efforts predominantly on the first two areas is logical, of course, since improvements in these two areas to some extent obviate the need for attention to the second two.

Our present state of knowledge and ability to estimate the effectiveness, benefits, and costs of the many recent highway and vehicle safety standards and those soon to be imposed can only lead one to doubt the necessity and desirability of some measures and to be hopeful about others. Among the more interesting measures currently receiving attention (and not previously discussed here) is speed control, either by speed-limit regulation or by speed governors on vehicles. Some limited data gathered in a few states and abroad suggest that the severity of accidents and perhaps the accident rate can be reduced by the use of more restrictive speed controls.[27] Severity reduction is easier to argue than accident-rate reduction, if one assumes that speed distribution does not become more skewed after a speed-limit reduction. The arguments for the former are straightforward: With a lower average speed, the momentum and energy to be absorbed in crashes will be reduced, on the average. Also, a lower speed limit should bring about a reduction in the relative speeds or speed differentials that exist in the flow, thus reducing the momentum and energy to be absorbed in crashes. Of course, the latter argument is a

---

[27]See, for example, *Research On Road Safety*, Road Research Laboratory, Her Majesty's Stationery Office, London, 1963.

major one in making assumptions about the effect on the accident rate; lower speed differentials should—everything else being equal—lead to reduced accident rates, since more time would be available to avoid crashes. At the same time, one should consider other possible effects of reducing speed limits and average speeds. For instance, if one takes seriously the *presumed* U-shaped cause-and-effect relationship between accident rate and vehicular speed, then one can assume that speed-limit reductions will reduce accidents in certain instances but increase them in others.[28] Presumably, at high speeds drivers pay close attention because of the obvious risks, and at low speeds drivers pay relatively little attention because of a presumed confidence and lack of risk and because there is time to react to and avoid crash situations. Such reasoning would tend to suggest a U-shaped relationship, since at very high speeds drivers would be unable to avoid potential crashes regardless of their alertness to the driving task. At the same time, some doubt must be cast upon this line of reasoning in view of the previously cited Road Research Laboratory study; specifically, its findings were:

If cars are grouped according to their observed speed relative to neighboring cars in the traffic stream, there is a U-shaped variation of accident rate (over a two-year period) with relative speed. That is, the average-speed vehicle has the lowest rate and the rates for very slow vehicles and for very fast vehicles are high.

The fact that "fast" and "slow" drivers have high accident rates does not in itself necessarily mean that relatively high and low speeds are the only direct cause of the situation. It may imply that driving speed is also correlated with certain other driver traits which are likely to cause accidents.[29]

Finally, in commenting upon the desirability of imposing more restrictive speed limits—should it be proved that either the severity or the accident rate can be reduced by so doing—the effects on tire wear, fuel and oil consumption, repairs, and so on, as well as the effects on the value of higher speeds, need to be considered and integrated alongside any safety improvements. As noted earlier, drivers probably do not travel at higher speeds in ignorance of the increased risks of injury and death but in spite of them and because of a presumed higher value (to them) of added speed or reduced travel time.

A number of possibilities regarding vehicle technology appear promising, some in the near term and others farther in the future. Worth mentioning are:

Airbag restraint systems or pivoting-seat systems (such as the Kinematic Safety Seat or "Protect-O-Matic" system)

[28]See D. Solomon, *Accidents on Main Rural Highways, Related to Speed, Driver and Vehicle*, U.S. Department of Commerce, July 1964; also, see comments in J. M. Munden, *The Relation Between a Driver's Speed and His Accident Rate*, Road Research Laboratory Report LR 88, 1967.

[29]Munden, *op. cit.*, p. 13.

Energy-absorbing bumpers or vehicle frames

Energy-absorbing median or roadside barriers (such as rosebushes or other forgiving structures)

Automated highway systems

The first three fall into the near-term class, and the last is clearly a possibility having implications for the distant future. The first two alternatives are particularly appealing because of the obvious incentives for private industrial concerns to conduct, on their own, the necessary research and development for practical application of the devices, assuming, of course, that they are potentially marketable. However, it must be recognized that the present federal standards program invoked by the National Highway Safety Bureau does serve to inhibit these types of innovations in a serious and identifiable fashion.[30] Without significant changes in the National Highway Safety Bureau standards procedures, the requisite analysis and evaluation probably must be conducted by that agency, if it is done at all. The possibility of automated highway systems involves issues considerably broader than safety objectives and must find support in agencies having a broader charter than the National Highway Safety Bureau. The fragmentation of ownership, operation, maintenance, and control of highway transportation—among private firms and individuals and federal, state, and local public-works and regulatory agencies—makes it difficult to expect private organizations, or even any one governmental agency, to undertake even analyzing and evaluating this proposal, much less attempting to develop an operational system. The obvious agency for mounting any significant effort on automotive highway systems is the Federal Bureau of Public Roads, which has already sponsored a limited amount of work in this area.

The third area of safety measures and programs pertains to the rehabilitation of crash victims, once injured—to "emergency medical treatment."[31] As progress is made in the highway- and vehicle-safety areas that impinge on the so-called "first and second collisions," the need for programs in emergency medical care or rehabilitation effort will be at least partially abated. Since most of our attention and effort is, and probably will continue to be, concentrated there, the payoffs from programs in emergency medical care can only be viewed with pessimism. With over two-thirds of the fatally injured automobile occupants suffering injuries to two or more areas of their body, what would be the effect of saving their lives, assuming this were possible through

---

[30]Carlin, *op. cit.*, pp. 2 ff.

[31]Detailed information and knowledge pertaining to emergency medical treatment is summarized in H. M. Mitchell, *Emergency Medical Care and Traffic Fatalities*, RM-5637-DOT, The RAND Corp., 1968; however, the bulk of the data relates only to the care of accident victims who are fatally injured.

better emergency medical care? Would they be maimed to the point where they regarded living as less than worthwhile? Would their salvaged condition impose impossible hardships on their families and friends? These aspects have, to my knowledge, received virtually no analysis or consideration (in terms of condition after salvation). As for the practicality of instituting broad-scale emergency medical programs capable of providing fast, efficient service to virtually all traffic victims, wherever they may be on our 3.7-million-mile road and street system (the great majority of which is rural and spread over 3.6 million square miles of land area), we can be less than optimistic. In rural areas, vast resources would be required to provide emergency medical care anywhere nearly equivalent to that now available in urban areas. (Recall that over two-thirds of the traffic fatalities occur in rural areas.) Couple this with estimates of the salvageability of rural victims, and an even more dismal picture emerges. Waller estimates that at the present time, for rural victims of fatal accidents, only 13 percent are probably salvageable, 31 percent are possibly salvageable, 28 percent are probably not salvageable, and 28 percent have unknown salvageability.[32] Despite this expressed pessimism (on my part), Mitchell feels that the problem is sufficiently hopeful to recommend the following:

(a) More detailed research on medical description of traffic casualty from scene of accident forward.

(b) Investigate costs of various programs such as first-aid training, helicopter ambulances, etc., and get data to estimate their effect on the injured group.

(c) Large scale statistical studies on the surviving group regarding residual impairment.[33]

Safety measures and programs pertaining to the compensation of victims or their survivors for losses from accidents[34] can serve a dual purpose: compensation of accident victims for damage or income (or pleasure) forgone by them or their survivors; and provision of incentives for drivers to develop efficient driving habits and adopt proper amounts of safety (with respect to safety devices, and so on). Present-day insurance or compensation programs have several obvious disadvantages: Compensation generally depends on judgments regarding the guilty driver and the extent of the insurance he carries; compensation payments are often delayed, particularly when court action is necessary; costs for administering automobile insurance, both to the companies

---

[32] These data were taken from Figure 4, H. H. Mitchell, RM-5637-DOT, *op. cit.*, and were based on a California study by J. Waller and D. I. Manheimer, "Traffic Deaths. A Preliminary Study of Urban and Rural Fatalities in California," *Calif. Med.* 101:272, 1964.

[33] H. H. Mitchell, RM-5637-DOT, *op. cit.*, p. 28.

[34] See additional comments on this subject in Carlin, *op. cit.*, pp. 19 ff.

and to the courts and state agencies, are high; and insurance premiums are seldom differentiated properly, with respect either to the individual's driving ability and the safety of his vehicle or to the injury potential of all drivers (for example, with respect to the extent to which they use seat belts). To circumvent these and other problems of the present insurance and compensation system, a number of proposed changes have been receiving increased attention from the public press and from Congress. The most widely publicized insurance plan to alleviate the above problems is the so-called "basic protection" plan proposed by Robert Keeton and Jeffrey O'Connell.[35] The primary features of this plan are that:

Car owners or drivers would insure themselves and their passengers (i.e., they would be self-insured) against personal injury or damage, and would be compensated for their injuries or damages *up to certain specified limits* by the driver's own insurance company (in much the same fashion as present-day collision-insurance policies), regardless of which driver (if any) was at fault.[36]

Self-insurance *up to certain specified limits* would be compulsory.

All self-insured persons would be exempt from legal liability for negligent driving up to the specified compulsory insurance limits.

Such a system would have the advantages of reducing the delays until compensation payments are received, of providing compensation to the injured or their survivors (whether the driver was guilty or not), and of reducing the costs of administering auto insurance. Also, better financial incentives would be afforded with respect to "defensive driving" and equipping one's auto with safety devices. On the other hand, by eliminating the necessity (in a certain sense) for determining the responsible or guilty party, it seems clear that information on driving ability will be reduced in quality and that it will be increasingly difficult to differentiate between good and bad drivers. As a consequence, such a program may tend to push us toward average-cost pricing, toward cross-subsidies from good to bad drivers, and perhaps toward a generally careless attitude about driving. (If, in determining the compensation for damages or lost wages, it doesn't matter who was driving irresponsibly, there would appear to be less incentive for good driving with a self-insured plan than with the present insurance setup, when the responsible party pays all.) All in all, it looks as though one can argue realistically that external costs

---

[35] R. Keeton and J. O'Connell, *Basic Protection for the Traffic Victim: A Blueprint for Reforming Automobile Insurance*, Little, Brown, Boston, 1965.

[36] Note that this "self-insurance" policy differs from present liability insurance policies in that the latter compensate *other* than the insured driver whenever the insured driver is at fault, while the former compensates the insured driver and his passengers directly, regardless of which driver (if any) was at fault.

would be reflected to an even lesser extent than at present and that bad drivers could then afford to carry insurance and drive (where now they cannot). It is difficult to balance the pros and cons and to judge the overall wisdom of one plan or another. It is perhaps more sensible to suggest that the advantages of self-insurance plans are sufficiently interesting to warrant more complete analysis and evaluation. (Hopefully, the present insurance study being conducted by the Department of Transportation will provide some useful guidance in this respect.)

In closing, some comment is warranted about the state of our knowledge and about the analyses underlying our traffic-safety planning and design decisions. First, our past study and research scarcely permit us to model the driver–vehicle–roadway–environment system reliably and to understand the effects of different safety actions.[37] The system complexity, the lack of appropriate data, and the unavailability of research funds all contribute to the paucity of information and explanatory behavioral models. (Of course, one should ask whether large-scale investments in researching and modeling the traffic-safety system are justified, all things considered. The answer is far less than clear.) Second, most traffic-safety planning and design decisions rely largely (and sometimes wholly) on the judgment or whim of the individual having the design or decision responsibility. Often the new design feature is included or imposed simply because "someone" thinks it would be "good" for the traveling public or because the public "ought" to have it or use it. But seldom are comprehensive system analyses—which account quantitatively for the system effects and for the benefits and costs stemming from a planning or design change—made an integral part of the safety add-on or safety-standard decisions. As a result, we have few useful guidelines for assessing the validity of decisions in the past (and continued today) or for appraising the currently talked-about design changes. Nor do we know for sure or even vaguely whether our highways, our vehicles, or our control devices provide too much or too little safety. Until we can say with assurance that a "traffic-safety crisis" does exist and that some action does offer hope for bettering the public interest—all things considered and not simply "Is it safer?"—not only should we look at the wisdom of adding more safety features, improvements, and standards, but we should consider as well the wisdom of revoking or deleting previously established safety additions or standards.

---

[37]For a closer look at this aspect, see B. F. Goeller, *Modelling the Traffic-Safety System*, RM-5633-DOT, The RAND Corporation, April 1968.

A VIEW OF SAFETY
IN THE PETROLEUM INDUSTRY

DONALD L. KATZ

Modern technology, with its increased size and degree of sophistication, has been able to maintain a gradually improving record of safety—reduction of fatal and disabling injuries—by a concerted effort. However, accidents still occur, and they are quickly brought to the public's attention by our modern communication processes, which cause the public to be more aware of these accidents than ever before. The question is raised as to which direction an engineer should move in response to the general desire to improve our safety record.

Before looking for the answers to our questions, it is important that we consider our present state. Industrial concerns generally give equal weight to employee safety and public safety. They recognize that safety is a system, engineering design being only one factor therein. The nation carries out a tremendous effort in the field of safety, the development of codes and standards, applying such information in the design and construction of facilities, and in training personnel.[1, 2, 3] Assuming that the engineer is willing to increase the tempo of his activities to make our industrial complexes freer from accidents, what direction should he take? Although economics is an item not to be overlooked, how direct is it as a trade-off for safety?

---

[1] USASI *Operating Procedures*, United States of America Standards Institute, New York, 1967.

[2] H. H. Fawcett and W. S. Wood, *Safety and Accident Prevention in Chemical Operations*, Interscience Publishers, New York, 1965.

[3] W. Haddon, Jr., *et al.*, *Accident Research: Methods and Approaches*, Harper & Row, New York, 1964.

Public Safety versus Employee Safety

Personal injury is of great concern in the design, construction, and operation of plants and industrial facilities. Because of the proximity of employees to the installations, they have a greater risk should an accident occur. Industrial organizations are aware that the effectiveness of their personnel depends upon the safety standards imposed at all levels. Employees are rightfully intolerant of unsafe practices and will not remain long with an organization that does not show true concern for their welfare.

Although employees are closer at hand when accidents occur, they can be trained in the use of safety protection equipment to respond to emergencies so as to reduce casualties. To be sure, those employees who are involved in taking corrective measures in an emergency may run added risk over other employees or the public in the same way that city fire department personnel take a greater risk in case of fire than the general public does. Any catastrophic accident reaching the public beyond the space separation would involve many employees and thus be of equal concern to them.

Further, it can be pointed out that incidents that bring danger to people are likely to interrupt operations and bring property damage. These interruptions are vital to the economics of the industrial installation, so there is an overlap of interest in safety: i.e., the unsafe plant is likely to be unprofitable as well.

The Safety Record of the Petroleum Industry

The American Petroleum Institute (API) survey of the petroleum industry, with 184 oil and gas companies reporting, shows 74 fatal accidents in 1968 (Table 1). Some accident facts reported by the National Safety Council are given in Table 2, and comparisons are made in Table 3. On an absolute basis,

TABLE 1   Petroleum Industry Fatal Injuries, 1966–1968.[a]

Field	1968	1967	1966
Refining	15	18	14
Marketing	22	16	10
Production	13	15	14
Pipeline	12	10	13
Not described	12	7	27
Total Cases Reported	74	66	78

[a]Source: American Petroleum Industry.

TABLE 2   Accident Facts (U.S.), 1967[a]

Total deaths	112,000
Principal classes of accidents:	
Motor vehicle	53,100
Work	14,200
Manufacturing	1,900
Construction	2,700
Public[b]	20,000
Railroad	700
Air transport	1,500
Fires, burns	1,000

[a]Source: *Accident Facts*, National Safety Council, Chicago, 1968.
[b]Excludes motor vehicle and work accidents in public places. Includes recreation (swimming, hunting, etc.), transportation except motor vehicle, public building accidents, etc.

all fatality records leave much to be desired, and the petroleum industry is no exception. On a comparative basis, the petroleum industry, which handles combustible material, including gases under pressure, has a record a bit better than for all workers and is better than agriculture and the construction industry by a factor of around four.

TABLE 3   Comparisons of U.S. Fatalities and Injuries for 1967

Field	FATALITIES (Employees per Fatal Injuries)	DISABLING INJURIES (Injuries per 1,000 Employees)
Petroleum industry[a]	$\dfrac{370,364}{66} = 5,600$	$\dfrac{5,555}{370} = 15.1$
All workers[b]	$\dfrac{74,000,000}{14,200} = 5,200$	$\dfrac{2,200,000}{74,700} = 29.5$
Manufacturing[b]	$\dfrac{19,000,000}{1,900} = 10,000$	$\dfrac{480,000}{19,000} = 25.2$
Construction[b]	$\dfrac{3,800,000}{2,700} = 1,400$	$\dfrac{230,000}{3,800} = 60.5$
Agriculture[b]	$\dfrac{4,000,000}{2,700} = 1,480$	$\dfrac{230,000}{4,000} = 57.5$
Street and highway workers in 1961[c]	$\dfrac{455,667}{211} = 2,150$	$\dfrac{21,104}{455} = 46.4$

[a]Source: *Review of Fatal Injuries in the Petroleum Industry for 1967* and *Annual Summary of Injuries in the Petroleum Industry*, American Petroleum Institute, Washington, D.C., 1968.
[b]Source: *Accident Facts*, National Safety Council, Chicago, 1968.
[c]Source: *Work Injuries and Work Injury Rates in the Street and Highway Department Industry*, Report 296, U.S. Bureau of Labor Statistics, Washington, D.C., 1965.

In 1967, a segment of the petroleum industry had 19 on-the-job fatalities; the same group of employees (87,924) experienced 41 off-duty fatal accidents.[4] Such information and a study of *Accident Facts*[5] soon leads one to the conclusion that workers in the petroleum industry are as safe on their jobs as when off duty and that they enjoy considerable safety advantages over many other workers even though they handle potentially hazardous substances.

The above records do not report the degree of involvement of the public with industrial incidents. The descriptions of the 1967 fatalities do not indicate that anyone other than the employees was injured in the same incidents. A study of the natural-gas-pipeline industry showed over a 15-year period that there were 34 fatal injuries to employees and 31 to the public, an average of about two persons per year in each category.[6] Generally, petroleum manufacturing plant accidents do not involve the public. Major incidents, especially those releasing vapors, could spill beyond the confines of the plant. However, concern for employees represents the best interests of the public as well.

We are concerned here with the engineer and his part in safe designs. A review of the fatal cases shows that only a small fraction of the accidents can be attributed to inadequate design of major equipment. It is true that ever larger and more sophisticated units are being built and that the potential for major accidents is increasing. Let us take a brief view of the current efforts to design, construct, and operate safely through developing codes and standards. Educational efforts of engineers and operating personnel are a part of the on-going program to make the industry safer for employees and the public.

Design Safety Managed through Standards and Codes

Safety is a prime consideration in organizational activities involving codes and standards. Four organizations stand out as developers of consensus standards for safety. These are:

United States of America Standards Institute (USASI)
National Fire Protection Association (NFPA)
American Society for Testing Materials (ASTM)
American Society of Mechanical Engineering (ASME)

---

[4]*Review of Fatal Injuries in the Petroleum Industry for 1967*, American Petroleum Institute, Washington, D.C., 1968.

[5]*Accident Facts*, National Safety Council, Chicago, 1968.

[6]*Summary of Safety Practices of Natural Gas Companies*, Independent Natural Gas Association of America, Washington, D.C., 1966.

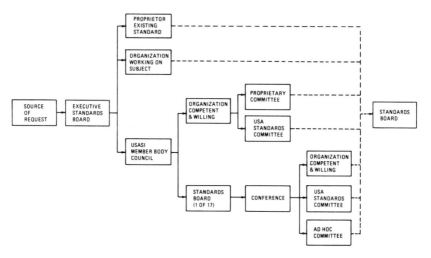

FIGURE 1   Flow chart–development of USA standards from USASI operating procedures. (From USASI *Operating Procedures*, United States of America Standards Institute, New York, 1967).

The USASI is a national coordinating organization representing industry, consumers, and government that establishes voluntary national standards.[7] It is a federation of 150 trade associations, has 800 company members, and includes a Consumer Council. As a capping agency, it has approved over 3,000 U.S. Standards. Figure 1 shows how other associations bring codes and their standards to the USASI for consensus approval. Table 4 lists some organizations of interest to the petroleum industry that sponsor standards development through the USASI or provide proprietary standards.

The preparation of a standard may be considered a pooling of the knowledge and experience of those who work in the area of concern, along with engineering considerations. Collective judgments as to adequacy of designs and good practices are involved. When incidents have occurred, the information derived from examining the incidents is utilized in review of the specifications or code. Part of the standards-preparation procedure is to prepare carefully worded statements that will accurately convey the information to anyone wishing to use the standard. The use of standards in design is very helpful in that governmental organizations on the federal, state, and local level can adopt them knowing that the best thinking to date has been incorporated in the procedures they are accepting.

[7] USASI *Operating Procedures*, United States of America Standards Institute, New York, 1967.

TABLE 4   Some Organizations That Develop and/or Sponsor Safety Standards

American Industrial Hygiene Association
American Concrete Institute
American Gas Association
American Institute of Architects
Institute of Electrical and Electronics Engineers (IEEE)
American Petroleum Institute
American Society of Civil Engineers
American Society of Safety Engineers
American Society of Heating, Refrigerating, and Air-Conditioning Engineers
American Society of Mechanical Engineers
American Society for Testing and Materials
American Welding Society
American Water Works Association
Compressed Gas Association
Cooling Tower Institute
Manufacturing Chemists Association
National Bureau of Standards
National Fire Protection Association
National Safety Council
Society of Automotive Engineers
Tubular Exchanger Manufacturers Association
Underwriters' Laboratories
U.S. Departments of Labor, Interior, Transportation, and Health, Education, and Welfare

Among the engineering societies, the ASME is a leader in developing safety codes and standards. Figure 2 shows their organization devoted to safety matters.

Some idea of the magnitude of the efforts involved in preparing standards, codes, and specifications may be gleaned from the fact that a 187-page book is devoted to listing information sources of standards and specifications.[8] Likewise, the index list of ASTM standards fills 153 pages; the standards fill 32 volumes. The NFPA publishes 10 volumes comprising 7,000 pages, including 34 USA standards.

The design engineering group for a company develops for its organization a list of standards and codes to follow—the index list of standards could run to 25 pages or more. The standards themselves represent a small library. Each organization normally adds or superimposes its own standards when experiences indicate the need. Generally, design standards relate to a specific item—a pressure vessel, a fire hose, an electrical device, properties of a steel, a flange design, for instance. Essentially all components of a plant have some standard or code that is followed in the design.

[8]E. J. Struglia, *Standards and Specifications Information Sources*, Gale Research Company, Detroit, 1965.

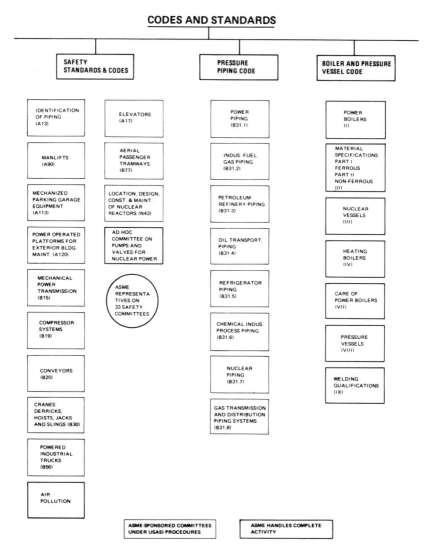

FIGURE 2   Organization of the American Society of Mechanical Engineers as it pertains to safety matters. (From ASME, *Organization, Activities*, New York.)

Education for Safety

The training for safety and communication of information is an important aspect of safety. The American Petroleum Institute's safety committee dates from 1920, and it provides manuals and guides. The National Safety Council has petroleum and chemical divisions. It is recognized that human failures, normally due to inadequate training or discipline, are responsible for a major share of accidents. In-house training programs are very thorough and effective in petroleum organizations.

Society and association committees interchange information on incidents as they arise, gaining knowledge of improved practices or the need for up-graded codes and standards. The series of programs of the American Institute of Chemical Engineers on safety in process plants showed great interest in this subject among chemical engineers.[9]

The relatively fine safety record of the petroleum industry is a measure of the educational and communication efforts currently being carried on as well as of efforts for safe design.

Economics and Safety

Few instances can be found in which there is a direct trade-off between safety and economics, although in the long run, suggestions for increasing the safety of a plant normally involve more expenditures. Take the example of a pipeline carrying natural gas cross-country. Here a given safety factor is established that is more than adequate when the metal is sound, laid properly, well pro-tected from corrosion, and used at the specified pressure level. Should corro-sion occur, no one can deny that a thicker-walled pipe might have delayed a leak that would be discovered and repaired or might have prevented a rupture that can be caused by corrosion. Here, to take corrective measures, one must decide whether money might not be better spent to remove the probability of a corrosive condition than to increase the pipe weight, which otherwise is adequate.

Starting with a given set of codes or specifications and raising the question of what could be done to reduce the frequency of accidents and improve safety, one needs to look for the weakest link and spend the money in strengthening it. Frequently, the finger is pointed at an item that, if improved by costly measures, might not improve safety because it is not the weakest link.

Which way should engineers direct their effort to fulfill their responsibili-ties to the public?

WHAT TO DO NOW

Knowing that:

The petroleum industry has a broad on-going safety program that has produced a good record;

[9]*Safety in Air and Ammonia Plants*, CEP Technical Manual, Vol. 1–10, American Institute of Chemical Engineers, New York, 1967.

In spite of this effort, some 70 fatalities of employees occur per year;

The nation is becoming ever more conscious of industrial accidents through modern communication channels and ever more impatient with our progress in improving safety;

The industry is building ever larger units and using more sophisticated plants,

how can the engineer make added contributions to better protect the public? If one were going to allocate more resources of time and money to safety, what should be done?

Four items appear worthy of consideration as answers to the above questions:

Support well the ongoing standards, safety-code, and educational activities that undergird our safe practices today.

Stimulate the study of safety as a total system.

Develop experimental programs to test the lesser-known design performances or weaker links.

Develop overall safety-analysis procedures.

The necessity for continued and enlarged support for code- and safety-standard-making organizations seems evident, based on past performance. Our nation has an effective way of obtaining a consensus for appropriate standards to safeguard the public. Likewise, continued safety training of employees is imperative.

The other three items, though not necessarily novel, do require discussion.

Consideration of Total System for Safety   The engineer knows that in addition to design considerations a host of other factors make up the total safety of a plant. Since factors other than the design of components are often responsible for failures that cause accidents, attention to the total system is needed both in the design stage and in the operation.

Codes and standards are preponderantly concerned with components of a plant, e.g., the pressure vessel, the piping, the electrical fixture. Seldom does the standard refer to the other components of the system that might have an influence on the individual component. An illustration will be given with which the writer is familiar in which the system concept has been adopted—relief-valve design. Valves are sized to provide a given flow of fluid under specified pressure conditions, generally with critical flow in the nozzle of the valve. Work during the past 25 years in connection with relief valves has been concerned not so much with the valve itself but with the many and diverse mechanisms by which generating vapor may be relieved. Also, back-pressure effects on the valve when it vents through a stack are of concern. Discussions of relief valves in refineries now refer to the pressure-relieving system as a whole, as in the recommended practice.[10] Codes or standards generally refer to a device like the relief valve and stop there. In this case, the relationship to the remainder of the plant became clear and the system concept was used in considering relief-valve practices.

Evidence that safety is recognized as a result of the total system is available. Sommer's paper discusses the elements of the total system, including both the design and the operating aspects.[11] His outline is given in Table 5. Levens' paper on systems analysis as a tool for accident prevention is noted.[12]

Through reports on accidents and conversations with persons experienced in management of safety, one learns that the "unforeseen circumstances" that are often responsible for accidents might have been avoided by a review of the total system. Such a review should become part of the design and should be made at intervals for operating plants. Teams representing such interests as design, construction, operation, emergency-handling, employee training, systems analysis, and management should review the safety of the total system. Corrective measures should be taken when any interaction of equipment and people has a given probability of creating a hazardous condition.

Computer simulation of plant start-up, operational upsets by disturbances, and shutdown is a way of evaluating the expected performance of a process plant.

Measurements to Evaluate Weakest Links   The weakest links in engineering design are worthy of study and experimentation. With the long-standing emphasis on codes for individual units, it could be that the greatest unknowns are in the interrelationships of assembled components during start-up and operation. Evaluation by plant measurements of such items as stresses, pressures, temperature gradients, and cyclic processes to find whether behavior matches assumptions in design is one way to search for weak links. Again, the emphasis is placed on the influence of one unit or another as representing the "unforeseen." The design engineer and the analysis team for total system safety are in the best position to nominate areas in which measurements or tests would be helpful.

[10]*Recommended Practice for the Design and Installation of Pressure-Relieving Systems in Refineries*, API RP520, 3rd ed., American Petroleum Institute, New York, 1967.
[11]E. Sommer, "Accident and Fire Prevention and Protection," unpublished paper, Esso Research and Engineering Company, 1966.
[12]E. Levens, "Systems Analysis, a Powerful Tool for Accident Prevention," *Trans. Nat. Safety Counc. 19*:25 (Petroleum Section) (1967).

TABLE 5   Some Components of Refinery-Accident- and Fire-Prevention System[a]

SAFETY OF PROCESS

Chemical reactions
Explosive compounds formed
Runaway temperature or pressure
Toxicity
Pilot plant or operating experience

PROCESS DESIGN

Temperature and pressure
Sequence of operations
Type of product storage
Overpressure protection
Instrumentation and alarms
Water, steam, electricity failure
Closed valve, instrument failure
Back flow control
Disposal of wastes
Knockout drums

EQUIPMENT DESIGN AND CONSTRUCTION

Materials, corrosion, attack
Brittle fracture
Wind, vibration, lightning
Pressure vessel codes
Electrical equipment
Equipment-industry standards
Basic practices
Weak roof seams

FIRE CONTAINMENT

Layout and spacing
Access and firebreaks
Fireproofing, insulation
Material, cast iron, brass
Curbs, dikes, sewers
Water sprays, monitors
Isolation valving

FIRE PROTECTION

Extinguishers
Water
Foam
Communications, alarms
Mutual aid

OPERATION AND MAINTENANCE

Operator knowledge
Turnarounds
Metal inspection
Visual inspections
Instrumentation checks
Start-up, shutdown
Work permits
Visitors and contractors

PROGRAMS, TOOLS

Initial and follow-up training
Accident and fire prevention
Radioactive isotopes
Fire-fighting training
Safe operations committee
Simulated emergency
Worldwide experience

[a]Source: E. Sommer, "Accident and Fire Prevention and Protection," unpublished paper, Esso Research and Engineering Company, 1966.

Overall Safety-Analysis Procedures   Acceptance of the value of system analysis and review for improving safety brings an assignment to the doorstep of the standards- and code-developing organizations. Surely the sharing of experiences and communication of analysis and review procedures to the industry is worthwhile within the framework used in the preparation of standards and codes. The products of such studies could be recommended practices and procedures.

In the final analysis, whether to make some modification to improve the safety of a plant depends upon an evaluation of the probability of an accident without the change. Reviews of industry-wide accidents and near misses as revealed in safety-team analysis of operations should help to set quantitative levels of probability for which corrective action is indicated.

How are safety decisions made? First, one follows all codes and standards that generally represent collective quantitative measurements and judgments. Then for the remainder of the design and interrelationship of components, one uses his judgment and experience. Efforts in the future should look to bringing collective judgments and measurements to bear on the total system, including the human factors.

Acknowledgments

Acknowledgment is hereby made of the assistance given to the author by many people, including E. H. Young, J. S. Queener, R. S. Sommer, R. B. McKee, R. W. Lindgren, T. Miller, and S. McCutcheon.

THE CLOSE RELATIONSHIP BETWEEN EMPLOYEE SAFETY AND PUBLIC SAFETY IN MODERN CHEMICAL PLANT DESIGN

J. SHARP QUEENER

After 40 years of experience in the chemical industry, the last 19 years involving safety, insurance, fire, and plant protection responsibilities, I asked myself, "Does the chemical industry consider the matter of public safety in facility design?" My first inclination was to answer in the negative because the Du Pont Company is so oriented to safety for the individual employee that the other side of the coin—safety for the public—is not as obviously apparent. On further reflection, I quickly realized the correct answer is just the opposite. The industry has its obligations to the public very much in mind when it designs facilities. The industry's reputation can have a very important effect on its ability to purchase property on which manufacturing facilities are to be located, to secure personnel, to transport intermediate and finished materials, and to sell its products, so the answer becomes very obvious.

Discussion in this paper will be directed to exposure of the public to danger arising from manufacturing and storage operations. Transportation and product safety matters will not be considered at this time.

When viewed in this context, has the chemical industry exposed the public to any serious degree in the past? What have been the forces or pollutants responsible for presenting simultaneous hazards to large numbers of the public, or what great potential damage has the industry posed to the public? A survey of case histories in publications of the National Fire Protection Association, the American Insurance Association, the Manufacturing Chemists Association, and others, which include fires, explosions, and toxic releases (to air, ground, or water), does not present a very significant picture as far as personal injuries to members of the general public are concerned. Yes, there have been some. Possibly the area of highest potential risk involves releases of toxic materials, particularly to the air.

The point I especially want to make is that in designing and operating chemical facilities, it is absolutely essential to consider both employee safety and public safety and that a creditable job has been done to date on both counts.

One of the most important public-relations challenges confronting the chemical industry is that of the public's attitude toward the relative safety of chemical-production facilities. Some years ago a nationwide attitude survey completed for the industry corroborated this fact: The public tends to believe that the chemical industry is considerably more hazardous to workers and to plant communities than are most other industries. This opinion is held generally in spite of the fact that the industry's record of safety achievement is among the very best in the nation. Not only has it been important to establish good safety performance, but an educational program has been found necessary in the communities where the industry operates in order to build confidence in the industry's safety. Experience has shown that any mishandling of an emergency can quickly destroy confidence, and competent emergency planning is therefore a necessary part in building public confidence.

On the whole, chemical-industry exposure of the public to personal injury from operational mishaps has been minimal. There have been instances of fume releases, such as chlorine and ammonia, that have caused evacuation of sections of communities, but serious injuries resulting from fires, explosions, and pressure ruptures over the last decade have been relatively few (less than ten fatalities). Most cases of this type of loss have been limited to property damage.

The in-plant safety performance of the industry has been very good—twice as good as the average experience of all industry. Latest figures of the National Safety Council show an "all-industry" frequency rate of disabling injuries per million man-hours worked of 7.22, compared to a rate of 3.55 for the "chemical industry." The severity of the injuries in the chemical industry was one-third less than for all industry. Some of the best occupational-safety performances in American industry have been achieved by chemical companies. An article in the January 1967 issue of the British magazine, *Industrial Safety*, entitled "The Truth about Work Safety in U.S.A.," says that the average frequency rate (ratio of injuries to hours worked) in the chemical industry in the United States is more than seven times better than that achieved by similar firms in the United Kingdom.

Importantly, the reason behind this good performance emanates from the Manufacturing Chemists Association (MCA), whose membership accounts for at least 90 percent of the country's chemical production.

In 1945, the MCA organized a General Safety Committee (now the Safety and Fire Protection Committee). One of its most important functions was to help prevent serious chemical disasters that could reflect adversely on the entire industry. From the work of this committee have come chemical-safety data sheets, safety guides, chemcards for transportation safety, a book and film on "Safety in the Chemical Laboratory," chemical-safety workshops held throughout the country on a geographical basis, and an industry-safety-award plan. Several other committees of the M C A have also engaged in safety activities, such as one that developed a labeling and protective-information guide for product packages.

Other associations have done much to promote chemical safety, such as the American Institute of Chemical Engineers (A IChE) and the American Chemical Society. At the March A IChE meeting in New Orleans, there were six sessions devoted to loss prevention in chemical manufacturing and storage operations. This topic is included on the agenda annually. Subjects discussed this year included risk evaluation of chemical plants, hazard classification of chemicals, fire research, anatomy of a nitration explosion, estimation of area affected by a toxic-fume (chlorine) release, explosion protection, and controlling large losses in storage facilities. The National Fire Protection Association and the National Safety Council have very active and strong chemical sections as part of their activities.

Because I am much more familiar with Du Pont experience than that of other chemical concerns, I would like to speak of some company information pertinent to the subject. In reviewing losses to the public sustained from operational occurrences, I could find no record of injuries to members of the public. There were of course some property-damage losses, but they were minimal. In a recent 10-year-period loss study, less than 2 percent of the total losses were in the public-liability category.

In a recent 12-year study of submajor and major types of occupational injuries within Du Pont relating to basic injury causes, such as method of doing, nature of process, lack of safety state of mind on the part of the employee, insufficient training, applicable safety device not provided, it was found that inadequate or improper design of machinery and equipment accounted for less than 7 percent of the injury total. Fundamental personal failures accounted for over 75 percent of the cases. These findings tend to verify the quality of engineering design effected in one company.

An explanation of an approach to facility design and determination of pertinent safety parameters can be further illustrative. The primary responsibility for obtaining necessary process information, including all safety aspects, lies with laboratory and pilot-plant group leaders and their technical people. This responsibility is an integral part of their jobs and cannot be delegated to others. The responsibility begins as soon as work is started on the process, whether on laboratory, pre-pilot, or pilot-plant scale. Then the process developed in the research laboratory is carefully checked by a task force committee. The hazard characteristics of both in-process chemicals and the finished product are considered. A study may be initiated in the Du Pont Haskell Laboratory for Toxicology and Industrial Medicine or in its explosives-research laboratory. Whatever the hazard, design is developed to contain it. At this time attention is primarily directed to employee safety, but public safety actually is an integral part of the consideration.

During the war, I was working on an ordnance plant Du Pont was building for the federal government in Memphis. Personnel procedures involved a suggestion system, and usable suggestions were given monetary recognition. Out of a box one day came the suggestion that a lot of injuries could be prevented if, instead of building a new dynamite plant, the dynamite needed was just bought. So far, this parameter has not been one suitable for design recognition by engineers.

Intensive study is made to find a suitable location for a new facility and enough property to permit a buffer zone. This provides protection from the possible exposure to hazards by other facilities and makes exposure of others less likely. Usually this buffer is helpful in case of unplanned fume releases. There must be adequate water supplies for both process and fire control. In fire-protection design, Du Pont has established, as have several other chemical companies, a standard of a maximum probable loss of $1 million, using separation, fire walls, sprinklers, and other devices to attain the limitation. Observance of nationally accepted standards is basic, although for special reasons they may be exceeded. Close attention is given to waste disposal. Deep disposal wells that discharge several thousand feet below ground level are provided in some locations; retention basins are frequently provided; blower-driven exhausts discharging at sufficiently high elevations to prevent dangerous concentrations at inhabited levels have proven effective; extremely pungent odors are vented to condensers that entrap vapors that might create a community nuisance; and in some cases, waste acids are transported by barge over 40 miles out to sea for slow discharge.

For years, attention has been given to prevention of air and water pollution. At one location, storage for 20,000 tons of ammonia was provided in an underground cavern. At another, a large ammonia-storage tank is of double-wall construction with nitrogen pressure between walls. In these instances, public safety was a prime concern. In air pollution, the possibility of exposing the public to any high noise emission is also guarded against.

Yet attention to design safety and the safest design possible will not ensure public safety unless safe operation is effected. Even the use of limiting-effects mechanisms, such as quick-closing valves and fail-safe devices, cannot do the whole job. There must be a safe operating philosophy. There must be adequate training, efficient preventive maintenance, periodic test procedures, well-trained fire brigades, process-hazard reviews, and preferably an outside-the-plant (not necessarily outside the company) safety and fire-protection audit on a periodic schedule.

Of proven importance to chemical plants is an emergency plan detailed for every location. Nonscheduled drills help to verify effectiveness. Membership of a facility in a community mutual-aid group has proved advisable. It has been demonstrated over and over again that the public has been better protected in times of emergencies with this type of preparation and organization.

In an attempt to decide how much safety is needed in design, an engineer finds there are many factors to consider—economic, psychological, and social. Even though the competitive picture and technical discoveries have created the need for higher-volume storage, higher temperatures, and higher pressures, it has been because of traditional concern for employee safety, the elimination of risk to capital investment, and the continuity of operations that economic considerations rarely endanger public safety. The possibility of a "trade-off" between economic gain and public safety becomes moot.

The chemical industry has been logical and foresighted in following its general approach in designing for employee safety, because in effect the public interest has been automatically represented. The approach therefore integrates employee safety with public safety. The imposition of risk or acceptance of high risk is not germane. There is improvement still to be made, but the direction has been effective.

A continuation of this design policy by the chemical industry is consequently well justified and should adequately protect the public interest in the future.

SAFETY AND RELIABILITY OF
LARGE ENGINEERING STRUCTURES

ALFRED M. FREUDENTHAL

Design at present of a structure for operation in the future on the basis of past experience necessarily embodies predictions of expected load patterns and load intensities as well as of expected structural performance. By their very nature, such predictions, which provide the assumptions and criteria of the design, must be made in the face of the uncertainty characteristic of any prediction based on past experience or on extrapolation from previous observations. If the uncertainty arises solely from the random nature of the phenomena involved, it can be dealt with rationally by probabilistic reasoning and, possibly, by the application of statistical methods which, therefore, must become an integral part of the procedures of structural design and analysis; any conceivable condition of the structure is associated with a numerical measure of the probability of its occurrence.

The predictive use, in structural design and analysis, of the theory of probability implies that the designer, on the basis of his professional competence, is able to draw valid conclusions from the probability figures obtained, so as to justify design decisions that, in most cases, hinge on considerations of economy or of social utility. It is not implied that this use is in itself sufficient to make a design more reliable or more economical, any more than the avoidance of the probabilistic approach makes it safer.

An approach based on the direct specification of a very low failure probability alone suffers from a major shortcoming: There is no intrinsic significance to a particular failure probability. Not only does its selection remain an arbitrary decision, but the use of a uniform probability level throughout a design is not necessarily a desirable feature, since the failure of different parts of a structure usually has quite different consequences in terms of the performance of the structure as a whole.

It is at this point that criteria must be formulated with whose aid the arbi-

82

trariness can be resolved. Such criteria are associated either with economic success in terms of expected financial gain or return on investment or with some measure of social utility or effectiveness that is to be maximized. Alternatively, it may be associated with economic failure in terms of expected loss, which is to be minimized by the selected probability of failure of the design. Different economic criteria or different definitions of utility will result in different optimal levels of structural reliability. While the selection of a suitable criterion for the design decision is still arbitrary, such arbitrariness is only that always associated with the choice of any economic criterion or criteria for social action. Its justification can be attempted within the framework of decision theory by balancing potential benefits against potential cost. The resulting "decision rule," by determining optimal or acceptable probability levels for the safety and reliability analysis of the structure to be designed, provides the link between its load analysis, stress and deformation analysis, and failure analysis.

In current design procedures the three aspects of analysis are practically uncorrelated, with no attempt made to assess the level of safety or the associated probability of failure of a structural design. In fact, only a vanishing number of our universities even attempt to make the engineering student aware that safety analysis exists and has a rational basis that permits the correlation of safety and probability of failure. We teach stress and deformation analysis as if it were the only aspect of structural analysis, without warning the student that the relevance of the most sophisticated stress analysis depends essentially on the relevance of the load analysis underlying it, as well as on the use that can be made of its result in the subsequent failure and safety analysis. The results of the most elaborate numerical analysis of a complex system made possible by computer development are largely wasted if they do not form part of an integrated load–stress–deformation–failure analysis and safety or reliability analysis, but are combined with more or less arbitrarily established "criteria," such as "load factors" and "safety factors" or "allowable stresses" that are specified by the codes assumed to represent the "public interest."

It is frequently implied that the safety provided by the codes is "absolute." This is far from the truth, even if we disregard the notoriously unpredictable intensity of seismic effects. Thus, for instance, analysis of structures for which wind represents a critical loading condition has revealed probabilities of failure associated with designs based on current criteria of loading and allowable stress of the order of 10^{-3} and higher.

In these and similar simple cases in which the dynamic interaction of load and structural response can usually be neglected, the problem of safety analysis is schematically illustrated by Figure 1. According to the statistical theory of extreme values, the extreme load intensity S increases with its return period $T(S)$ according to the expression

$$S = u_s + \alpha^{-1} \ln T(S), \tag{1}$$

where u_s denotes the most probable load intensity and α a parameter that is inversely proportional to the scatter range of load intensities. The resistance R either may be constant or may decrease with time (corrosion) or with the number of load cycles (fatigue).

In the first case, failure by pure chance will occur when the safety factor $v = R/S$, which is a statistical variable obtained as the quotient of the variables R and S, of distribution $P(v)$ and density function $p(v)$ assumes values $v < 1$. Therefore, its probability under a single load application:

$$p_F = P(v)_{v=1} = {_0}\!\int^1 p(v)dv = 1/T_F, \tag{2}$$

where T_F denotes the return period or mean time to (chance) failure.

In the second case the time to failure is a statistical variable with a mean and a range indicated on Figure 1 and expresses the joint effect of gradual reduction of resistance and increase of extreme load intensity with time. The probability of failure thus becomes a function of time that can also be related to the momentary safety factor $v(t) = R(t)/S(t)$, which defines a family of distribution functions $P\{[v(t)]\}$ as well as the related probabilities of failure:

$$p_F(t) = P[v(t)]_{v=1}. \tag{3}$$

In the less simple cases in which the interaction between the local and the

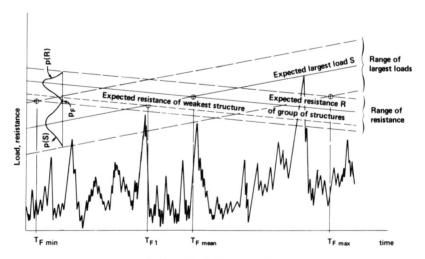

FIGURE 1   Safety analysis.

dynamic response of the structure cannot be neglected, the "local intensity" S to be used in the evaluation of $P(\nu)$ is the effect of the load S "filtered" by the dynamic response of the elastic structure. Under the simplifying conditions of "narrow band" filter action that is frequently satisfied, a relation similar to Eq. (1) can be derived for the extreme load effects, so that the scheme of Figure 1 still applies.

A rational approach has thus been outlined by which the probability of failure p_F associated with certain values of a mean or median or modal safety factor $\bar{\nu} = \bar{R}/\bar{S}$ or $\breve{\nu} = \breve{R}/\breve{S}$ or $\tilde{\nu} = \tilde{R}/\tilde{S}$ can be determined. The problem now arises of selecting a value p_F associated with a central ratio ν that satisfies realistic public-safety requirements within the framework of economic design. In such a design the functional and structural requirements have to be satisfied within the framework of an economic model that permits the construction of a measure of performance used as a criterion for the selection of the "optimal" structure. The problem type encountered is designated in decision theory as a "risk situation" in which, for each "course of action" defined by a certain structural design, the probabilities of the alternative "outcomes" of failure or survival can be quantitatively estimated on a rational basis. The objective of the "decision rule," by which the most desirable design (course of action) is to be identified, is to maximize its expected "utility" while minimizing the expected loss in the case of its failure.

It is, in general, impossible to set up a decision rule that would achieve both objectives simultaneously. In fact, it is not a priori certain that such a course of action exists. A common practice in this situation is to minimize the expected loss in the case of failure, while imposing a certain limiting condition on the utility. This formulation assumes that a course of action in which the imposed condition is not satisfied has no utility, while all other courses of action have the same utility. By repeating this procedure by varying the functional limitation while minimizing the expected loss in the case of failure, a set of designs can be obtained with corresponding "failure-loss-minimizing" resistance. The utility of the designs for different functional limitations must then be assessed and a design selected by engineering judgment, probably at a point where the increase of utility becomes marginal. Obviously, this course of action may not yield the design of maximum utility, but since it is associated with a design that minimizes the expected failure loss, it represents probably as good an approximation of an optimal solution as can be attained.

Alternatively, utility and expected loss can be combined additively in a measure of utility equal to the (negative) expected total cost on the basis of the assumption that the higher the utility the lower the total cost, combining the cost of construction with the expected cost of failure of the structure and its replacement, including the cost of loss of function during the period of replacement. Since the construction cost and therefore the cost of replace-

ment after failure are functions of the probability of failure on which the
design is being based, the objective of attaining maximum utility formulated
as the minimum expected total cost will produce relations between the functional limitations of the structure and the probability of its failure that can be
used in selecting an optimal design. This definition of utility produces, in fact,
the widely used decision criterion of minimum expected loss, since this loss is
represented by the expected total cost.

In order to apply this criterion, a set of structures must be designed that
are reasonably similar in appearance, satisfy the intended function equally
well, and would produce identical consequences when failing; they differ, however, in initial cost and probability of failure under similar loading conditions.
Within this set, the range of variation of initial cost and of probability of failure
is usually moderate, since large changes involve, in general, considerable
changes in design and appearance. The above set of structures can be extended
to include structures of different functional limitations that therefore satisfy
their intended function with different degrees of effectiveness, but the failure
of which would still produce similar consequences. In this case the initial costs
and the replacement costs are functions of both the functional limitation and
the probability of failure associated with the design.

Using the criterion of minimum expected loss L for the optimization of the
probability of failure p_F, this criterion can be expressed in the simplest form:

$$L = A + I(p_F) + p_F F \rightarrow \min, \tag{4}$$

where A denotes that portion of the initial cost that is independent of p_F
while $I(p_F)$ denotes the part that depends on the probability level, and F
denotes the total cost of failure made up of the replacement cost of the structure and the circumstantial cost of failure. It is this part of the cost of failure
that is determined by the public aspects of its consequences and that therefore
may be much higher than the cost of replacement, even including the cost of
loss of function during the replacement period. Assuming, in first approximation, that the weak dependence on p_F of F through the replacement cost can
be neglected, the decision rule (Eq. 4) leads to the equation

$$\frac{dI(p_F)}{dp_F} + F = 0 \tag{5}$$

for the optimal value of p_F.

Almost no information exists concerning the form of the function $I(p_F)$.
The not-unreasonable assumption that I increases linearly with a decrease in

$\log p_F$, or

$$dI = -cd \log p_F, \qquad (6)$$

leads to the relation

$$I - I_0 = c \log p_F, \qquad (7)$$

where I_0 denotes the value of I for a limiting structure designed to fail with certainty ($p_F = 1$). Combining Eqs. (5) and (6), the condition is obtained.

$$p_F = \frac{c}{F}, \qquad (8)$$

where c denotes the increase of the cost I associated with the decrease of p_F by one order of magnitude, which is some fraction of I, probably not exceeding a few percent. The optimal design probability of failure is therefore a similar fraction of the ratio I/F, which determines the order of magnitude of this probability. The highest admissible optimal probability of failure for structural parts whose failure cost does not much exceed their replacement cost is therefore of the order of $p_F \sim 10^{-2}$. As the economic and, particularly, the public consequences of failure become more severe, its cost is rapidly rising. In the case of the almost purely economic consequences of failure of an offshore drilling platform during a storm (when the crew has been evacuated), a recent study suggests a ratio $I/F = 1/10$, so that a design probability of failure o/$p_F \sim 10^{-3}$ would be optimal. This value might be considered as the upper limit for structures whose safety is not only of economic but also of public interest, since as soon as the public interest is involved, the ratio F/I rises very fast with the extent of such interest.

It is frequently said that failure probabilities of the order of 10^{-5} or 10^{-6} can have no real meaning, since they are too far out of range of verification by actual data. This contention ignores the principal purpose of the probabilistic approach to structural design, which is not to provide values that have an absolute meaning, but rather to provide a scale on which the safety of different designs can be compared on a rational basis, once an analytical procedure based on a relevant probability model has been agreed upon, and on which therefore the degree of public safety can be assessed.

CIVIL STRUCTURES AND PUBLIC SAFETY: SAFETY IN DESIGN AND CONSTRUCTION OF CIVIL STRUCTURES

JOHN A. BLUME

I shall refer to so-called fixed or stationary civil engineering structures such as dams, bridges, and buildings. I won't go into nuclear power plants because I might invade the territory of the following authors. The term "so-called" fixed structures was used because we should look upon all structures as moving, vibrating, and responding to ground motion and to forces or loading. This disturbance or loading can be called a "demand" for the purpose of generalizing. I like to use two types of demand on structures; one, the demands of nature; and the other, the demands caused by man.

Let me quickly run through a list of some of the demands of nature on civil engineering structures: windstorms, aerodynamic amplification of wind, hurricanes, cyclones, tornadoes, missiles induced by tornado action and other wind action, rainfall, snow and ice, runoff, overtopping, flooding, landslide, avalanche, earth settlement or rupture, temperature variations that may affect the structure seriously, volcanic eruptions, ocean swell, ocean waves, tidal extremes, tsunami, seiche, and earthquake. Earthquake can also generate many secondary hazards, such as fire, panic, health hazards due to broken utility installations and sewage lines, and of course tsunami, seiche, and earth slides. Not all structures are subjected to all of these demands, although one would be surprised how many of them can affect a structure during its lifetime.

There has been a lot of publicity in recent months about California sliding into the sea. I did not list this as one of the demands for obvious reasons. Since this conference has as one of its main points working closer with the public on risk and safety problems, press coverage becomes important. I would like to inject a little side note here that in my experience and in my opinion, there has been more news-media coverage of California sliding into the sea in recent months than there has been of essentially all earthquake research in the last several decades. This is unfortunate.

Now let us consider the demands caused by man himself. There is first dead load, which is the simplest demand of all to determine. There is live load of various types and fluid pressure. There is vibration, impact due to collision, such as that of a ship on a dock or a wharf, or of an automobile on a barrier. We also have man-made ground motion nowadays from nuclear events; and we have air blast from explosions or sonic boom. And there are forces that may be created in soils or rock by excavation. Even fire is a demand that must be resisted, and so is water and other fluid erosion.

Natural demands are seldom under the control of the designer, except by judicious location of his structure. Very often he does not have any say as to where the structure is to go. So one might say that these natural demands are random variables that often vary over an extremely wide range, the mean values and the limits of which may not be fully known or recognized. Earthquake loading especially is highly skewed probabilistically, and it may have extreme high values with very low probability, values that could be several times the mean or median value. What to do in design about the low probabilities of very high natural demands is a serious problem and one in which we should involve the public.

Man-made demands may or may not be under the control of the designer or the analyst. In any event, most of them are also random variables without specific deterministic values. The random variables may be subject to appropriate probability laws. Often, however, data necessary to do a very good job in modeling the probabilities are lacking. I think it very important that the public somehow recognize, and the layman realize, that the engineer does not always know specifically what the real demands are going to be. He may have little real data with which to work. This is especially true in the earthquake field, to which I will slant my remarks today, although most of what I intend to say applies in general to the other demands as well.

Earthquakes are subject to much wider variations or extreme values than any other demand, with the possible exception of tornadoes. There may be a very, very small probability of something very severe happening, and yet it can, and does, sometimes happen.

Speaking of small probabilities happening, in California last winter there was very severe rainfall and runoff in many areas. One method of design is to use an average frequency for a certain intensity of demand as, for example, a 100-year flood, or a 100-year runoff, or a 500-year storm. This is not the same return method that Dr. Freudenthal was talking about. In early January there was about a 200-year frequency runoff in a certain area in California, and 30 days later there was another one of essentially equal intensity. So the probability of getting two 200-year demands in one month shows the trouble one can have with extreme demands and low probabilities. A "500-year" earthquake might occur tomorrow, or 1,000 years from now, and there is no way to know the difference. Our hindsight is very short, and yet we must have foresight. We lack sufficient data to forecast very well, but we face the problem in design for safety of what should be done now.

I would like to define the "capacity" of the civil engineering structure as its yield value, or the point at which distress first occurs. The distress might be stretching, or it might be actual failure. This depends upon the characteristics of the structure. We must consider capacity, of course, in the same unit as demand. This might be force. It could be acceleration, velocity, displacement, pressure or stress, or many other units. Capacity is to some extent subject to the designer's control, with, of course, economic restraints.

When demand exceeds capacity, when the ratio of demand over capacity is greater than one, damage begins. This is a joint probability if one assumes that demand and capacity are independent, and they usually are. For example, if the probability of demand were 10 percent and the probability of a certain capacity were 10 percent, the joint probability of these two getting together would be 1 percent, the product of the two. Whether or not this 1 percent is acceptable is another question.

The amount of damage depends in part on how much the demand exceeds the capacity. It also depends on the inelastic characteristics of the structure beyond its yield point. If the structure is brittle, yield and failure are reached essentially simultaneously. There is perhaps a loud bang and all is gone. This means collapse. If there is a ductile structure, it may stretch out, like taffy. It may not collapse. Of course, between these two idealized outer bounds there are all sorts of other combinations.

I look upon the energy-absorption capacity of structures under the severe demands of nature as the new dimension in civil engineering design. In addition to strength, there should be consideration of energy and work capacity. I know a few others in the National Academy of Engineering feel the same way.

Since we don't know exactly what the probabilities of demand, or even capacity, are, what do we do in design? Certainly, judgment has to enter the picture. There is one approach, if one wants to be more rigorous in the analysis, and that is to resort to a Bayesian type of philosophy. I know this is controversial in certain academic circles, but under this philosophy, one uses statistical models and procedures with all of the data one has, but professional judgment is introduced also to apply corrections within certain limits as one goes along. The final result is thus based upon a combination of data, mathematics, and the best available judgment. I think this approach has great possibilities in the field of design for public safety.

The consequence of damage is the really important matter to the public. What risk is really acceptable? I am always surprised at the amount and degree of surprise that follows every major earthquake. The newspaper headlines become about a foot high, especially if crystal-ball seismic predictions are also currently in the news. The public is shocked, and many engineers, even many who should not be, are also surprised. I don't share this surprise, because I have felt, written, and lectured for decades that the probability is small, but it is there. It is just a matter of when and where it happens. In Caracas in 1967 four buildings collapsed completely, and one collapsed partially; about 280 people died. It was not really a major earthquake. In 1968 there was a severe earthquake in Japan that essentially destroyed eight school buildings of three and four stories that had been designed to specifications that would pass in California today. This is going to be investigated in detail, we hope. We are working on this because it should be a valuable example to show that small-probability events do happen and can have serious consequences.

People were greatly surprised at what happened in Anchorage, Alaska, in 1964. Ever since Anchorage and the development of nuclear power plants in this country and elsewhere in the world, there has been something that I am very glad to see after all of this time—a renewed, and I hope a continuous, interest in the earthquake problem and what it can mean in public safety.

Perhaps we engineers should not try to determine the acceptable risk level all by ourselves. Perhaps we should consider some other factors and let the public participate. Certainly we have to consider the law, and the law changes from time to time. We have to consider the moral aspects. We have to consider the social aspects and the professional aspects. I think we have a responsibility to tell the layman what the problems are, but without scaring him to death needlessly. And, of course, the technical aspects must be considered.

Putting this another way, I feel the profession has a responsibility to protect the public against itself, in the earthquake field especially. Now, what do I mean by that odd statement? I mean that engineers are not financing and building buildings—the public is. The public is controlling the budget. Most designers are going blissfully ahead and designing buildings according to what-ever minimum demand requirements may be in effect; a few designers may go a little beyond that and be accused of being "strong." The public is interested in economy. This is a natural result of our operations in supply and demand and of our economic system. There is nothing wrong with the system, *per se*. However, when one is talking about the nebulous risk of something that may or may not happen 5, 10, or 50 years from now—the small probability of severe natural demands—it is pretty hard for the businessman to part with the extra dollar. This is especially so when he feels his competitors may discount the risk entirely.

The added cost must be weighed against the benefits. The benefits may only be decreased risk. I have found in many years of practice in designing struc-tures to be earthquake-resistant (we don't use the term "earthquake-proof" for a building) that the additional cost is often not very great, sometimes prac-tically nothing, providing, and only providing, the engineer takes an active part in the basic layout of the structure. If he is brought in by the architect, for example, after the basic layout and space allocation have been crystallized, the chances are he can't do very much toward creating real earthquake-resistance unless he spends a fair amount of the owner's money in additional construction costs.

I think architects should get into this picture more than they do. Most of them share the feeling of the public, and the layman, that there are building codes and these building codes should do the job. Others feel that if an engineer is on the job, the problem is automatically solved. Few engineers understand and accept probabilistic concepts. Many who do discount the risk or feel the restraints of the economic system.

The earthquake building codes are not intended by the people who write them to be a panacea or to solve all of the seismic problems. They are only minimum standards. The unfortunate part of economic life is that almost without exception minimum standards become the only standards. It is not intended in any existing earthquake code that I know of that there would be no damage after a severe earthquake in the area. Damage may in turn cause injury, or worse in rare cases. Another problem is that the building codes lag considerably behind knowledge. Whether this is a 5-year lag or a 10-year lag is hard to say. I could note examples of 10 years or more. Why does it take so long? I think you will find that all things of this nature take a long time. New findings have to be debated, tried, tested, and argued, and there must be public hearings. Often compromises or watered-down legal requirements result from this process.

Another situation is that the codes are based to a considerable extent upon the performance of traditional-type buildings, such as those with a steel or concrete frame almost entirely encased with "nonstructural" filler walls of masonry or concrete. These filler walls were not used in the computations, so the result was excess strength in the structures that caused them to have an earthquake performance that has been reflected in the building codes. Today these codes are being applied to contemporary buildings that often do not have filler walls and excess materials.

Essentially all of the lay public and an amazing number of engineers and scientists seem to feel that actual earthquake performance—good or bad—is the essential test of a structure or of a seismic-design method. The type of structure that has survived an earthquake seems to be acceptable. Although performance is important, as a sole criterion or as the only important one, it fails to define the earthquake or to recognize the great variations in the demand not only from different possible earthquakes but over the area of a city in the same earthquake. (We hope to demonstrate this soon from our work in response to ground motion from underground nuclear explosions.) It also fails to recognize the great variations in buildings and in soil conditions. A building may actually survive one earthquake and be a poor risk for other earthquakes. A soldier who goes into battle without getting wounded is not necessarily bulletproof. Some worthwhile aseismic advances have been tabled because no earthquake had "confirmed" the need or the theoretical findings. On this basis, we could never send man around the moon because there would be no precedent!

A dangerous concept that applies to very tall buildings is that if the wind forces and responses govern over the seismic forces and responses, the latter may be ignored in design. First of all, this fails to allow for the higher modes of seismic vibration of very tall buildings that may induce critical stresses. But above all, it does not allow for the greater extreme values of earthquake (as compared to wind) as related to the normal deterministic design values. Again, the probability may be low, but is it low enough to be ignored, and who decides this?

I use the term "vagrant architecture." I am not referring to the architect himself but to a type of building that is very popular today, especially on the West Coast. It is a building that looks as if it has no visible means of support. These are very popular on the covers of many architecture magazines, and the people like them. The designs win awards for their clean, daring appearance. The unfortunate part is that many of them truly have very little support for the case of a real emergency that might exceed by several times the design-code values. The probability may be low, but it exists.

I am a veteran of San Francisco's early attempt, a few years after the end of World War II, to adopt its first real earthquake code. It may surprise most of you to think that San Francisco didn't really have one before that time. Even then, it took several years to put it into effect. I think this is something we have to examine carefully. The public reacts against new codes or code increases. If we want more public safety, the answer is not simply to go out and say, "All right, let us improve our building codes. Let us, for example, take a code that calls for 5 percent base shear and make it 8 percent." It is not that easy. The minute you try to increase a building code in most local communities, and this applies to large cities as well, all sorts of people come to the meetings. The property owner, for example, who owns existing property designed under the 5 percent requirement says, "If you change that code to 8 percent, you are making my building a second-class investment, and I won't stand for it." Then he gathers all the other property owners, and they talk to the elected public representatives about it.

Labor unions have complained that the code may increase building cost so much as to prohibit future building. All of these things are factors that come into efforts to improve a building code for reduced risk. It is a problem that has to be faced. Perhaps the best approach would be to have a group such as the National Academy of Engineering, and other appropriate groups, take such a stand after real study of all the facts and risks as to practically force the issue into being if it is jointly agreed by everyone including the unbiased public that such should be the case. It is either that or wait for a major disaster, after which they often do change building codes, *in that particular area*!

One of the still-unrecognized real issues, even in the professional community and certainly on the part of the public, is that we have had an unbalanced state of technological development, one that is highly advanced in certain areas but woefully lagging in others. We have large fast computers and advanced theory that together can solve essentially any problem. As an example, for years a few of us have been subjecting models of large complex buildings and other structures to the complete time history of entire earthquakes; the output is the complete response of the model and all of its maximum forces and unit stresses under the earthquake demand. We can repeat this for any number or type of earthquakes. The results are no better, of course, than the ability to model the real structure or the ability to select the earthquake that will, or might, shake the structure in its lifetime.

How does one define "might" in that sentence? What probability is to be considered and what probability is to be ignored? In the nuclear field this varies with the structure and with the consequences of failure, as it should. In the normal building field, the approach is not this at all but merely "to meet the building code." In fact, because of the advances in technology and increased knowledge of materials as well as better materials, the tendency is to increase the unit stresses, use limit or ultimate design, reduce safety factors, and so on. This is fine if, and only if, there is full knowledge about the demands that may affect the structure in its lifetime; the probabilities of these demands and of the capacities to resist them; the consequences of the severe demands such as tornado or major earthquake; and an acceptance of the risk and of the consequences by the professionals and by the public. This phase of our technology trails other advances, and it may lag for some time into the future unless the old-fashioned deterministic approach to demand is changed.

It is the nature of man to take risks. This is how he has survived, how he has advanced, and often how he receives the acclaim of his fellows. But one can no longer risk the other person's money or life without his approval. A person in an automobile or a plane knows there is some risk, perhaps extremely small, but he accepts this for value received. Most people do not recognize any risk in a civil engineering structure such as a dam, a bridge, or a building. Usually this risk is extremely small, but some failures have occurred. More could occur in the future as the population increases and as more time elapses to permit the laws of probability to operate.

Let me make it clear that I am not saying there is a vast probability of cities falling down tomorrow. I am saying that there is some probability that some buildings may collapse in the United States in the future. The earthquake problem is not just in California. A new map shows it is everywhere in the country except parts of Florida and Texas and a small piece of Alabama and Mississippi. It varies across the country, but it is there based upon geologic findings, our short seismic history, and current activity. Certainly earthquakes, tornadoes, and hurricanes aren't going to stop. What is the solution to this situation?

It seems to me that we have to:

1. Continue to gather all the data we can on the actual demand, whether it be earthquake or other demands, and continuously record reliable data on what really occurs.

2. Make proper statistical models on what these demands really are.

3. Set by law (code) or by professional standards the minimum demands that must be designed-for in various areas based upon the risk and consequences of failure determined to be acceptable. This determination should somehow be made in conjunction with the public.

4. Stop oversimplifying dynamic design with static equivalents. Modern theory is advanced; large, fast computers can do wonderful jobs, and judgment can also be introduced. The day is coming when available new design methods may be appropriate. In addition, energy and work capacity, as well as strength, should be considered for structures.

5. Treat probabilistic problems as such, not with inapplicable deterministic approaches that fail to recognize possible severe variations in demand or capacity. This must be done especially if and where computer solutions are even partially substituted for the "art" of design as practiced in prior decades by experienced men with sound judgment and a sense of structural adequacy.

6. Re-educate designers for changing values and concepts or else provide them special help for special problems.

7. Put existing knowledge to work; knowledge today and for the last decade or more is far more advanced than most realize. However, carefully consider the limits of the data at hand.

8. Keep the public informed, but don't scare the people needlessly.

I have a simple expression, the word "CARE," which I believe applies to this safety problem: Compatability Analysis of Risk and Economy.

PROVIDING FOR PUBLIC SAFETY IN
THE NUCLEAR INDUSTRY —
THE ENGINEERING APPROACH

JAMES T. RAMEY

The topic of public safety as a growing factor in modern design is timely; there is a great need to focus on the problems that engineers face in the design of engineering systems having the potential for simultaneous hazards to large numbers of people. Although I was not technically trained, I have been around the atomic energy program a long time and have been accused of thinking like an engineer.

I will describe briefly the measures that have been taken to provide for public safety in the nuclear power industry, discuss some of our experiences, and stress the importance of a disciplined engineering approach to satisfy our safety requirements. These matters will be discussed in light of the unprecedented commitments that have been made by the utilities to nuclear power plants and the increasing importance of electric power to our overall health, safety, and well-being.

Meeting the safety requirements of nuclear power has always been recognized as a prerequisite to its successful application. From the beginning, the entire development of atomic energy has been characterized by a deep concern for the safety of the public, of those in laboratories and industry who carry out atomic energy programs, and of those on nuclear ships or otherwise involved in its military applications. This concern was perhaps best expressed by Chairman Holifield of the Joint Congressional Committee on Atomic Energy when he said:

The atomic energy program is unique in that for the first time a detailed regulatory system was imposed by the government before the experience of any serious accidents prompted a demand for such regulation.[1]

[1] In "Recent Developments and Future Challenges in Atomic Power," presented at the Edison Electric Institute Convention, San Francisco, California, June 8, 1966.

This policy of providing for public safety in advance of adverse experience is in contrast to safety policies and practices that have accompanied technology applications in many other areas. For example, the loss of the *Titanic* with 1,490 persons is a vivid instance of safety regulation coming only after a tragedy had occurred. Despite clear warnings over many years that regulations on subdividing ships into compartments were not adequate, effective steps to develop and apply subdivision standards were not taken until the *Titanic* went down. The history of technical advances contains many other examples of failure to provide adequately for safety in advance. On the other hand, in the development of atomic energy there has been an ever-present emphasis on seeing that safety requirements are anticipated and provided and that essential safety research and development is carried out.

Regulatory Review Process

It may be useful to describe briefly the regulatory process that is so essential to meeting this objective. While my remarks will relate to the regulation of commercial power reactors, the basic approach is also followed for other types of nuclear facilities. When the Atomic Energy Act was amended in 1954 to permit the private use of atomic energy, Congress provided for the federal regulation of atomic energy facilities "to protect the health and safety of the public." The regulatory program that evolved from this legislation has been an effective instrument in helping to assure the safe application of nuclear energy to civilian uses.

In order to build and operate a nuclear power plant, the utility owner must have a license. The first step in obtaining a license to construct a nuclear plant is usually an informal safety review by the Atomic Energy Commission (AEC) regulatory staff of the plant site proposed by the utility.

The next step is submission of the license application, including a detailed safety-analysis report, and its subsequent review by the regulatory staff. The objectives of this review are to obtain or develop adequate technical information on the reactor design and its site, to reach an understanding of the technical basis for the safety of the proposed plant, and to enable the regulatory staff to make an independent evaluation of the safety of the plant. In the safety analysis report, the applicant must describe the safety characteristics of the proposed site and reactor and review a wide variety of postulated equipment malfunctions, operating errors, and other unusual occurrences and analyze the capability of the proposed plant equipment to protect the public from adverse consequences of such events.

Nuclear power plants use a defense-in-depth concept to assure adequate protection of the public. There are three basic lines of defense:

The first and most important line of safety protection is the achievement of superior quality in design, construction, and operation of basic reactor systems so as to ensure a very low probability of malfunctions.

The second line of defense consists of the accident-prevention safety features that are designed into the plant. These systems are intended to prevent malfunctions of the reactor systems from escalating into more serious accidents. Such features include emergency reactor-shutdown systems, emergency cooling system for the reactor core, and emergency power systems.

The third line of defense consists of consequence-limiting safety features, such as containment. These features are designed to confine or minimize the escape of fission products if they should accidentally be released from the fuel and reactor systems.

The regulatory staff reviews the safety-analysis report submitted by the applicant in sufficient detail to reach an independent judgment that the safety of the reactor is adequate. As part of this review the staff must determine which extremely unlikely potentially serious accidents have sufficient probability of occurrence to require protecting the public against their consequences. The possible consequences of the most severe accident in this category are analyzed in considerable detail, and a judgment must be made that the maximum potential radiation exposures would not result in serious injury to the public. It should be emphasized that a reactor would not be considered acceptable by the AEC unless any accident that could result in such potential exposures is judged by the AEC reviewers to have an exceedingly low probability of occurrence.

A large spectrum of other possible accidents is also reviewed in detail from the standpoint of probability and potential consequences. If these accidents are judged to have a higher probability of occurrence than the severe accident referred to above, a judgment must be then reached that potential radiation exposures would be correspondingly less, in order for the reactor to be acceptable. In other words, in examining possible accidents, both the probability and potential consequences are considered in reaching a judgment concerning the acceptability of the risk involved. The judgments to which I refer are highly informed ones made by highly qualified technical staff. They require evidence of the most thorough engineering of the reactor and its systems, supported by extensive and intensive engineering reviews and accident analysis.

Continuing efforts are, of course, made to improve the methods used to evaluate safety. For instance, we have considered at various times whether a more quantitative approach than that described could be used to evaluate risks associated with nuclear power. We have requested an Internal Study Group, recently appointed by the Commission, to consider this matter as part of its study of the regulatory process. While the report of this Group has not been completed, we have been advised that the Group has concluded that with existing techniques and knowledge, the total risks to the public from nuclear power plants, although very small, cannot now be meaningfully expressed numerically. However, the Group indicates that these quantitative techniques show promise for use in making comparative safety evaluations and that efforts should be made to improve the collection of data needed to evaluate quantitatively the reliability and causes of failure in safety-related systems of nuclear plants.

In addition to the review by the Commission's regulatory staff described above, a further independent safety review is made by the Commission's Advisory Committee on Reactor Safeguards (ACRS). The ACRS, it should be noted, is made up of exceptionally qualified technical experts from outside the AEC. This Committee is required by law to review and report on each major power- and test-reactor application. The utility owner presents its safety evaluation to the ACRS, and both the utility and the reactor-plant supplier respond to questioning from the Committee. The ACRS provides recommendations for research-and-development efforts in these areas of concern. Both the ACRS report and a separate and complete evaluation of the safety of the proposed reactor by the regulatory staff are made available to the public.

The public hearing is the next step. The purposes of the hearing are to inform the public of the safety aspects of the proposed plant and to provide a means for any person whose interest may be affected by the construction and operation of the proposed plant to advocate or oppose the granting of the license to construct the plant. The hearing is conducted by the Commission's Atomic Safety and Licensing Board, composed of one lawyer and two persons with technical backgrounds. Statements are made early in the hearing by the applicant and the regulatory staff, describing in layman's language the steps that have been and will be taken to assure public safety. After hearing the views of all parties, the hearing board makes a decision, subject to Commission review, as to whether the license should be granted.

Only after satisfactory completion of each of these three technical reviews—those of the regulatory staff, the ACRS, and the licensing board—followed by a review of the record by the Commission, can a license for construction of a plant be granted. Similar reviews are also carried out before an operating license is issued, although a public hearing is not usually held.

The test of this regulatory process and of our safety policies is, of course, whether they have been successful. The record indicates that they have been. Since the beginning of the regulatory program, the Commission has licensed the operation of over 20 power reactors that have accumulated over 88 reactor-years of operating experience. No member of the public has been exposed to radiation levels above permissible annual limits as a result of operation of these licensed power reactors.

This favorable experience provides us with a measure of satisfaction and confidence. But it has not and must not lessen the emphasis on safety considerations. We must be constantly alert to the results of developing experience. Careful account must be taken of the problems being encountered. I would like now to mention some of these problems, particularly those related to technical considerations. To place the discussion in context and provide appropriate perspective, let me provide some background information on the developing application of nuclear power.

Growth of Nuclear Power

Beginning with pioneering efforts of the 1950's, the nuclear power industry underwent a sustained but modest growth until 1965. The year 1965 saw the start of a substantial commitment by utilities represented by the purchase of seven nuclear power plants. The commitment continued to grow, with 20 plants purchased in 1966, and 31 plants in 1967. In 1968, orders dropped somewhat to 17 plants because of a number of factors, including the traditional cyclic buying pattern of utilities. All told, there are now about 78 civilian power plants in the United States, either under construction or on order, with a projected capacity of 63,000 MW, in addition to those nuclear power plants that have already gone into operation. The AEC estimates that by 1980 the installed nuclear power generating capacity will be between 120,000 and 170,000 MW and will represent 25 to 30 percent of total U.S. electric power capacity.

These unprecedented commitments to nuclear power plants have important implications with respect to the ability to meet rapidly growing electric power needs. The reliance placed on these plants as dependable sources of power in the early 70's makes it most important that they be built on schedule and perform in a safe and reliable manner in accordance with prediction and plan.

Nuclear power plants have been and can be built to provide this kind of performance. Much experience has been gained from the successful design, construction, and operation of Shippingport, Yankee Rowe, Dresden, and other first-generation plants. Further important information is coming from second-generation plants like San Onofre and Connecticut Yankee in the 450- to 500-MW range. But it is also important to note the significant increase in the size of these units to the 1,000-MW units that have been ordered. There is a lack of experience with these larger-size plants and with the higher power densities and other engineering extrapolations involved. Also, we are mindful of the instances of deficiencies in the design, construction, testing, and operation of some Commission-owned and commercial reactors. Because they have been identified sufficiently early to take corrective action, these deficiencies have not resulted in safety problems. But many have been costly—often very costly—in time and money. Consequently, not all of the many nuclear power plants under construction or on order will come into operation on the originally estimated time schedules, with the performance and at the costs projected.

Two examples will illustrate the kinds of deficiencies that are of concern. In one instance, an AEC inspector detected cracks in the assembly casting of a pump that was being installed in a reactor under construction. Further investigation disclosed similar defects in all other pump assemblies of this kind. Corrective action resulted in the replacement of a substantial number of castings for pumps to be installed in several different reactors. In another instance, welding defects were found in the emergency core-cooling system of a reactor plant. Subsequent radiography disclosed that three-fourths of the welds in this system were unacceptable and required costly reworking. All defective welds were repaired and reradiographed and were then found to be acceptable.

Importance of Quality Assurance

Most of the deficiencies to which I refer have resulted from inadequate engineering. That is, they have been manifestations of failure to use a disciplined engineering approach and to apply it thoroughly and systematically throughout all phases of the design, construction, testing, and operation of nuclear power plants. Such a disciplined approach requires forceful quality assurance, understood as comprising all those planned and systematic actions necessary to provide confidence that a structure, system, component, or plant will perform satisfactorily in service. It includes but is not limited to quality control, which consists of those actions related to the physical characteristics of a material, component, or system and which provides a means to control quality to predetermined requirements.

There has been a growing awareness over the past few years that strong quality-assurance measures are essential to the safety of nuclear power plants. We believe that the Commission, its regulatory and development organizations, and the ACRS may take a fair measure of credit for bringing about this awareness. I have urged the need for quality assurance in many speeches and by other actions to all elements of the nuclear community—utilities, reactor plant suppliers, industry groups, and professional societies. I mention this because I consider that those of us in government with responsibilities for public safety must provide leadership consistent with these responsibilities.

Despite the growing awareness of the importance of quality assurance to safety, I consider that it needs to be increased still further. There is need for better understanding of the close relationship that exists between safety and reliability. We need to recognize more than we have that a nuclear plant that is capable of sustained regular operation, according to prediction and plan, has a high degree of inherent safety. I think it is not too harsh to suggest that engineers have not been sufficiently active in bringing about this understanding. I offer this suggestion in a friendly way, fully mindful of the many significant contributions that engineers have made in bringing atomic energy to its present state of development.

Better understanding of the close relationship between reliability and safety will have important advantages. It will help increase recognition that safety objectives are served first and foremost by designing, building, and operating nuclear power plants so that they perform as intended. It will result in greater attention to evidence of departures from normal conditions of reactors and to correcting incipient difficulties before they become more severe. It will also focus attention on the need for early detection of problems during construction and testing phases.

Recognizing the need for such priority attention, I have long urged that the greatest emphasis in reactor safety be given to a strong quality-assurance program throughout all phases of a nuclear plant project. This involves assuring that the design is adequate to meet defined and agreed-upon requirements, that construction is carried out in accordance with the design, and that tests confirm the design. It further requires assuring that the plant is operated with approved procedures by appropriately qualified operators within safe limits established in the design and safety analyses and that the plant is maintained in accordance with accepted and established procedures.

There is a particularly acute need for increased engineering attention and direction in the component fabrication and plant construction, test, and operation phases. Providing such attention and direction requires a break with traditional practices in certain kinds of construction activities where engineers are neither accustomed to providing nor required to provide procedures, instructions, and detailed direction of the kind needed in nuclear power plants.

It would be easier to bring about acceptance of the kind of engineering direction and control needed for nuclear power plants if there were recognized nuclear industry standards that set forth the requirements of an acceptable quality-assurance system and accepted implementing practices. A proposed addition to the Commission's regulations, "Quality Assurance Criteria for Nuclear Power Plants," has recently been approved for public comment. These criteria, developed by the Commission's staff, will contribute to meeting the need in this area. In developing these criteria, account was taken of cooperative AEC-industry efforts on establishing quality-assurance requirements and of the experience accumulated in designing, constructing, and operating licensed nuclear power plants and Commission-owned reactors. The proposed criteria will assist applicants or licensees by presenting in detail the Commission's requirements with respect to quality-assurance programs. I consider these criteria a very important step forward.

Earlier, I referred to the deficiencies that have been encountered in carrying out nuclear plant projects. These deficiencies have led to concern about the unwarranted cost increases and schedule delays that have resulted. But this concern has not yet brought about sufficient recognition of the savings in time and cost that can be made if resources are applied to doing a job right the first time. Extra costs can almost always be avoided by insisting on more engineering attention and more adequate quality assurance. While costs associated with delays and down-time are affected by many factors, they are usually formidable. It may cost $1 million for every month of delay in placing a 1,000-MW nuclear plant in service and up to $4 million for every month of down-time after that.

In order to better acquaint top management in the nuclear industry with the urgent need and importance of quality assurance in bringing nuclear plants into operation, the Commission is organizing a series of meetings with the industry. Specific problems and associated quality-assurance shortcomings in the Commission's programs and in commercial plants will be discussed, as will the quality-assurance criteria recently published by the Commission.

Role of Engineering Societies

The complexity and magnitude of the issues related to introducing advanced technological developments such as nuclear power involve a large number of organizations representing the various interests, both public and private, through which the engineering profession works. Foremost among these various interests and organizations are the engineering professional societies. The engineering professional societies provide the structure, leadership, and knowledgeable personnel to establish standards, to feed back experience and identify weaknesses, and to assist the engineering profession in best focusing its collective resources on the tremendous challenges and responsibilities presented by nuclear power.

I am convinced that we have the basic organizational structures and technical capabilities for establishing and meeting the exacting standards of safe, reliable, and economical nuclear power. However, all segments of the nuclear industry must recognize more clearly that they have a high-priority interest in supporting the standards activities of the engineering professional societies, in providing the leadership and knowledgeable personnel, and in underwriting the costs.

The procedures associated with developing engineering codes and standards, and their incorporation into mandatory requirements, are traditionally time-consuming. These traditional delays in developing and adopting codes and standards are further compounded by the long lead times associated with construction of nuclear power plants. Consequently, the results of current efforts in developing codes and standards will not be seen in operating plants for many years—unless special actions are taken to see that they are. Let me illustrate this with reference to the Nuclear Piping Code, which was published last year.

In 1957 the B31 Standards Committee on the Code for Pressure Piping sponsored by the American Society of Mechanical Engineers (ASME) established a subcommittee on nuclear piping to explore the special requirements for nuclear installations. Development of this code was a difficult job, requiring considerable effort. In February 1968, the ASME published a draft Nuclear Piping Code, B31.7, for trial use and comment. Comments are now being assembled, and a redraft of the code is being prepared for approval of the various code groups involved. It is expected that this code will be adopted and published later this year. Meanwhile, a large number of nuclear power plants are being built, and new orders are being placed. If the traditional approach should continue to be used, one would not expect to obtain the full benefits of this code in operating plants for perhaps another 5 years—15 years after the need for a nuclear piping code was recognized.

It is clear from this example that the traditional approach to the development, adoption, and use of improved engineering standards is not commensurate with the rapidly developing pace of technological developments such as nuclear power. Actions to correct this situation must be taken, both individually and collectively, by the members of the engineering professions concerned through the professional engineering societies and private organizations involved. Because of the serious implications involved, the Commission is considering what regulatory actions it should take in response to this situation and others like it.

Actions by the engineering profession must involve reconciling various complex interests, both private and public. For example, the difficulties in developing standards derive from many factors, including the procedures of professional-society standards committees, which require near-unanimous agreement to establish standards or to modify them. I recognize that the members of the standards committees reflect the interests of their employers. But they must also reflect the interests and public trust of the engineering profession. The point is that the standards committees of the professional societies should provide the forum to reconcile these legitimate interests in accordance with the traditional principles of responsibility that the engineering profession is expected to observe. These standards committees cannot be, as they sometimes have been, the means by which needed standards are blocked or delayed in the committee. Neither should the resulting standards reflect only the lowest common denominator of agreement. The members of these committees, as well as their employers and others whose interests they represent, must recognize that the issues involved must be faced and resolved responsibly.

In discussing the role of the engineering profession and professional societies in developing industry codes and standards for nuclear application, I would be remiss if I did not acknowledge the dedicated efforts of many individuals who have made substantial contributions in this area. Individuals like Murray Joslin, chairman of the United States of America Standards Institute (USASI) committee for developing standards for reactor plants and their maintenance, and Frank Williams, former chairman of the ASME Boiler & Pressure Vessel Code Committee, have provided important leadership in these efforts. But I am confident that they would agree with me that such efforts need to be greatly intensified.

Conclusion

Over many years, the Commission has given careful attention to fostering the growth and development of safe and economic nuclear power. As commitments to nuclear power grew to their present large proportions, we have been constantly alert to developing experience—both prospects and problems. Although our safety record is good, we have discerned a clear need for more thorough and exacting engineering measures of the kind I have discussed. The responsibility for meeting this need falls on many parties and organizations. Most importantly, it falls on engineers and the engineering profession. I am confident they will not fail the challenge.

REACTOR SAFETY:
THE CARROT OR THE STICK?

F. REGINALD FARMER

The operation of nuclear plants has, in the past, led to the escape of some fission products and doubtless will again. The release may be small, representing "no undue hazard," or it can be very large.

It is not then a question of whether reactors are safe, but rather of "How safe are reactors?" and "How safe should they be?" An answer "completely safe" is impossible, and "as safe as possible" is no answer, as further effort and money can reduce risk but might more effectively be applied elsewhere. Today we do not know whether the answer to the question "How safe are they?" lies close to the target that might be set or falls short by orders of magnitude.

These questions are not hypothetical, and the problems they pose are real, for if there is a shortfall in achievement compared with target, we may not know until too late, when our countries' commitment to the nuclear program leads us once again to pay the price for engineering advancement.

Profiting from the lessons of the Windscale accident in 1957, the U.K. Atomic Energy Authority formed a section to study reactor safety as an engineering and scientific discipline, rating this as an exercise in the development of atomic energy requiring staff of at least equal merit to those involved in research and design. With this section I have been closely associated with nearly every type of reactor system—gas-cooled reactors with magnox canned fuel; advanced gas-cooled reactors with oxide fuel; pressurized water reactors used for defense; research reactors; the sodium-cooled fast reactor at Dounreay; the high-temperature helium-cooled reactor (DRAGON); and the steam-generating heavy-water reactor.

Ten years ago we thought that safety might be enhanced by codes of practice, recommendations, safety principles, and criteria. We applied considerable effort to this work and found that, if developed unilaterally, the criteria were mostly wishful thinking and exhortations that were little used or were in such general form, as others I have seen, that it was easy for the user to claim conformity. Alternatively, if developed jointly with industry, the process of codification became lengthy and, for good reasons, an exercise in compromise more relevant to reactors built last year than to new types for the future.

I tried then to answer the question, "What standard is required for reactors of the future to give freedom in siting—to give low risk to the community?" and of course came back to the problems of how to measure safety and how safe society wants to be.

The potential harm from the release of fission products is an increase in the risk that already exists from natural causes of development of thyroid carcinoma, lung cancer, or leukemia. The measure of increased risk depends on the quantity and nature of the fission products released, the weather, the surrounding population distribution, and the biological relationship between dose and effect. Three of these four components are reasonably well known in comparison with the greater uncertainty of the fourth—the accident potential of the reactor. Nevertheless, we can set up a target that stems from a social objective and is independent of the reactor and then consider the impact of a particular target for the reactor designer.

An early attempt to set a target sought to fix an upper limit on the release of fission products—as in siting criteria—but this only transfers the problem to the measure of credibility or credulity and requires "reasonable assurance" that the release will be only one part in 1,000 to one part in 10,000 of that contained in the reactor. The designer has had no measure of credibility to guide him and he has his own standard for what is reasonable.

It seems sensible, and indeed obvious, to start with a requirement that as the potential release gets larger, the chance that it will occur be made smaller. This could be defined as a limit line, as in Figure 1, that requires that the chance of releasing 1,000 curies of iodine-131 (or other fission products of equivalent hazard) should for any reactor be less than one in a thousand per year, and for larger quantities a lower frequency. The implications of this choice will be seen later. All that is now implied is that fission products may be accidentally released from any one of the many reactors installed; that the release of concern may lie within one of the four decades 10^3 to 10^7 curies; and that it should be the designer's intent to make the likelihood of the release less frequent than that indicated on the line drawn.

We can for each decade of release apply the information on downwind diffusion from the known frequency of weather patterns, and then for various site population densities we can estimate the risk of thyroid carcinoma (for ^{131}I) using the current estimate of 10 to 20 cases per 10^6 man-rem.

The results for typical sites ranging from remote to urban are shown on Figure 2. There is too little space to describe these in full; the material has been presented in other papers. We should consider the assumptions and the

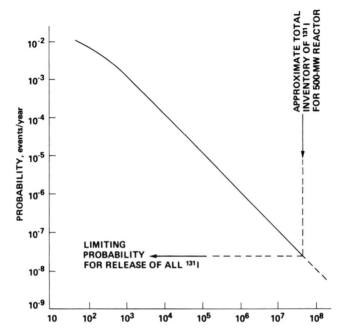

FIGURE 1   Probability limit line versus curies of iodine-131 released.

consequences. For an urban site—the upper line—the chance of some 30 cases of thyroid carcinoma occurring is less than 3×10^{-4} per year. This number, 30 cases, is barely detectable in a population of 300,000, which is the number of people in one downwind sector of the hypothetical site. If 100 reactors operate for 30 years, there is just a chance that the accident might occur once. This is a very small risk indeed, but I believe that the limit at this end of the scale is not fixed by harm to people but is more acutely felt as cost to the industry. Any small to medium accident will cause very substantial loss of revenue and repair charge, and if these occurred yearly, or every few years, mostly without harm to people, the target at the low-release end of the scale might be set even tighter.

How to move from a low-release target to a high-release?

We could compare events of recent history causing harm to few or many people; we may conclude that events involving many people occur less frequently than events affecting few. There are various reasons—weather, crowd distributions, and so on—that ensure that high penalties occur less often than low, but there is no law that regulates this. I believe that the exercise of retrospective analysis is not particularly helpful, and it is today a question of sensibly interpreting the objectives of society and government.

Let us try to exercise our own judgment.

Referring again to Figure 2, curves 2, 3, and 4 represent fairly typical semi-urban reactor sites, and these curves show that for the assumptions built into this exercise, there is a chance of less than one in a million per year per reactor that the number of cases of thyroid cancer could fall in the range of 1 to 10,000. Present-day decisions on reactor type and quality will affect some hundred reactors to which the country will be committed for 30 years. For this whole program, which is consequent on today's decisions, the chance of occurrence of an accident having this degree of severity is about 1 per 1,000. Perhaps this risk is pitched too low by comparison with the other hazards to which society is subjected, but I do not believe that we in the nuclear industry would receive a vote of confidence if we increased this risk to 1 in 10. It is possible that many would debate in the region of 1 in 100 to 1 in 1,000.

If you are convinced that our business should be run so as to keep risks down to this sort of level, then you are requiring reactors to be so designed

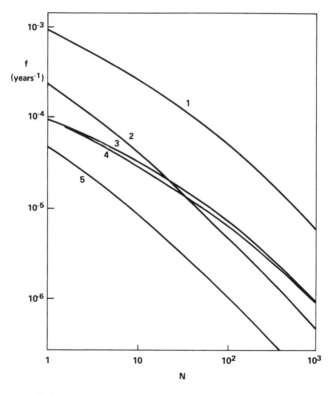

FIGURE 2   f-N curves for five population distributions.

and built by engineers (and well maintained and operated) that the likelihood of their accidents giving rise to fission-product release does not exceed the curie line previously shown. This requires some events to have a likelihood of less than one in a million per year.

There is a tendency to think of these numbers as associated with extremely remote events like meteorites and plane crashes on stadiums. But there are many events occurring annually on a reactor demanding shutdown action, water supply to boilers, or other routine functions for which a failure rate for the local system should be one in a million or less if their operation alone secures safety, or perhaps 1 in 10,000 to 1 in 100,000 if some allowance is made for the benefit of containment.

An engineer knows how to set about achieving this target. He will find out what information is available about the reliability of the type of equipment he proposes to use. He will find a scarcity of good information, but with persistence he will find a wealth of engineering experience on equipment of a very similar type. The use of this information will tell him simply and directly whether he can expect to meet his target using a suitably designed system made up of units conforming with the middle or upper range of present practice. If achieving this result requires an excessive overlay of redundant and diverse equipment, he may conclude that a new design is required for some items and will foresee the need for a development program and for providing trials. It is still likely that 90 percent of equipment will follow current design and manufacturing trends, but whether for new design or stock components, he, as the design engineer, can specify a performance requirement including a limiting failure rate.

Let me recapitulate the argument so far and the position now reached.

Reactor safety is not absolute but can only be seen as a graduated risk. Most of the risk stems from accidents and malfunctions of plant and misconceptions in design. Means to reduce risk taken by the designers will include a complexity of measuring and response equipment. To maintain a low-risk target, each system, subsystem, and component has a function to fulfill that can be specified, together with an upper limit of a tolerable failure rate. This can be done: A low-risk target can be achieved but it requires effort; it requires organization and incentive.

How and by whom can the work be done?

The assessment of the accident potential of the plant must fall to the designer. It is only he who knows the plant in detail and it is only his effort that can significantly change design and derive plant specifications. To complete his job, he requires more information on plant and component failures than he now has. Much of his plant is also used elsewhere, and information can be obtained from other industries and then, progressively, from the nuclear industry.

There are difficulties in obtaining plant information from operators. I believe also that there are many in the safety business who would prefer to guess and call it judgment. I regard it as extremely important for safety that the industry should proceed on the best information possible and then, and only then, use judgment; and if the best information is not available to it, steps should be taken to get the information and use it.

This leaves the choice of reward or directive—the carrot or the stick. It is possible that both may be required, but ultimately, work and effort are maintained only if there is a conviction that it is worthwhile.

Most people agree that systematic effort will increase plant availability, and in high-cost plants this is big money. It has been said that receiving and storing fault information is expensive—one system has been set up in the United Kingdom to receive and store information on more than 10,000 plant faults per year; the cost is estimated at less than a dollar per fault.

If costs are reasonable and the benefit large, why is effort of the type described not being applied systematically toward nuclear systems? As the effort is needed at the design stage, it is for the design organization to answer, but I hazard two possible reasons:

To date there has been inadequate conviction of urgent need; either what is being done is assumed to be good enough or all that can be done is being done.

The design organization (also manufacturers and utilities) no doubt fear that their judgment is challenged by safety organizations in a way that they feel is irrational. They then hesitate to commit themselves to a more disciplined and definitive effort to achieve a high safety target that is not so far officially accepted, fearing that the dispute with agencies and committees would lengthen the time scale for approval and incur additional plant penalties.

Is it likely that a considerable additional effort will be spent unless there is confidence of a common acceptance of judgment based on fact—experience reasonably assembled and interpreted? There are safety people who are suspicious of the overzealous use of factual experience that they feel may be used to cover a weak point; of course this will happen, and it may take years to build up confidence, but this will not be made easier by the preferred Delphic approach of some safety people—the oracle speaks but must not be challenged.

For many years, I have worked closely with an organization concerned with design, construction, and operation. There now exists a reasonably common understanding of a mutual objective: The objective of a graduated risk set at a low level has been approved by the United Kingdom Atomic Energy Authority, and the practice is being worked out. In most discussions on specific safety-design issues there is constructive argument, but agreement is reached on interpretation of experience and its application to the new plant within a reasonable margin. The engineer then goes ahead convinced of the rightness of a particular solution—convinced that it is his decision and not one arbitrarily imposed by a safety group.

You may feel that there are differences between systems and organizations in the United Kingdom and the United States. Of course there are, but not in the fundamental intent to succeed, and to succeed safely. You may feel that all my talk is really concerned with codes of practice and quality control. I agree that in part this is true, but only in part. A diligent development of such codes (if applied across the board) might improve engineering achievement by a small factor. This has been the whole trend of engineering history. It might decrease failure rate by a small factor. The future target for supersonic aircraft is, according to one report, to reduce the present failure rate by 30 percent.

I am talking about quality control in a very particular sense, not devised centrally, not devised to suit some hypothetical reactor, but generated by the designer to meet the specific requirements of the equipment for his design. Nobody else can do this job, and he cannot do the job unless he knows what standard of safety is required.

In conclusion, I advocate setting up a target objective for the nuclear industry, a meaningful objective capable of interpretation in engineering terms such as that given in Figure 1. The implementation will fall mainly on the design organization, who should be asked to present reasonable data to show how their plant might be commissioned, tested, and operated to meet the target. This essential information will then pass to the operator, who will set up an inspection and test schedule to ensure that the performance is consistent with that required by the safety standard.

CLOSING REMARKS

ERIC A. WALKER

Now we need to put a National Academy of Engineering frame around the picture we have seen painted. We have heard a number of excellent papers on the problem of public safety; many questions have been raised, and we have even heard answers for some of them. But the one question that remains for the Academy is, "Where do we go from here?"

Our pattern of operation has been to have a symposium on a subject that has an engineering backbone in which we think we might have some interest. Some of the Academy members will remember that we had a symposium on education and finally asked if the Academy should become involved in education. We have had other symposiums on traffic safety; ocean engineering; and science, engineering, and the city. Last year we discussed the engineer's role in medicine. The Academy now has some very active committees concerned with these topics.

We have talked about public safety, and clearly a good many engineers have given a great deal of thought to this subject. We have seen that there are different approaches in getting to the problem. But the need to represent the public interest and to evaluate questions of risk versus gain in the design of engineering systems is becoming increasingly important as the size of the systems gets bigger and the potential for damage gets larger.

There are many questions that we might ask: What role can the National Academy of Engineering play in guiding the approach to balancing risk versus gain in the design of large engineering works? What guidelines might be provided in answering the many ancillary questions involved in each design? What trade-offs are considered? How is reliability to be determined? What bases should be used to evaluate the cost–effectiveness of various safety features in terms of the reduced consequences, such as reduced property damage? How is the break-even point or optimization of safety to be identified? What guidelines are to be used in quantifying risks in terms of property damage, injuries, and fatalities that might result from postulated accidents?

One of our objectives in having a meeting like this is to determine the state of the art, how much we know about the problem, its susceptibility to an engineering approach, and how many other actors are already on the scene, because there is a limit to what we can do as an Academy. If this symposium has stimulated thinking on the subject, it has already served a useful purpose. However, should *more* be done by the Academy? For an answer to that, I must come back to the question that the N A E Council and members have to answer: Where do we go from here?

PARTICIPANTS

JOHN A. BLUME, President, John A. Blume and Associates, Engineers, San Francisco, California

JAMES M. BROWN, Professor of Law, The National Law Center, The George Washington University, Washington, D.C.

WILLIAM K. BYRD, Acting Deputy Director, Office of Hazardous Materials, U.S. Department of Transportation, Washington, D.C.

DAVID E. A. CARSON, Vice President and Actuary, The Hartford Insurance Group, Hartford, Connecticut

F. REGINALD FARMER, Head, Safeguards Division, Authority Health and Safety Branch, U.K. Atomic Energy Authority, Risley, Warrington, Lancashire, England

ALFRED M. FREUDENTHAL, Professor of Civil Engineering, Department of Engineering and Applied Science, The George Washington University, Washington, D.C.

DONALD L. KATZ, A. H. White University Professor of Chemical Engineering, The University of Michigan, Ann Arbor, Michigan

J. SHARP QUEENER, Manager, Employee Relations Department, Safety and Fire Protection Division, E. I. du Pont de Nemours & Co., Inc., Wilmington, Delaware

JAMES T. RAMEY, Commissioner, U.S. Atomic Energy Commission, Washington, D.C.

CHAUNCEY STARR, Dean, College of Engineering, University of California, Los Angeles, California

MARTIN WOHL, Manager, Transportation and Analysis, Transportation and Research Planning Office, The Ford Motor Company, Dearborn, Michigan

115